A Spectrum Through the Primes

A Concrete Tour of the Spectral and Quantum Approaches

Book Two of The Primes Trilogy

companion to *A Wheel Through the Primes*

John Maggio
Efficient Intelligence, LLC

Published by Efficient Intelligence, LLC

Paperback ISBN: 979-8-9967325-3-1
Hardcover ISBN: 979-8-9967325-4-8
Ebook ISBN: 979-8-9967325-5-5

First edition, 2026 Printed in the United States of America

Advance in Peace

Preface

This is the second of three books about the same set of facts. The first, *A Wheel Through the Primes*, built its picture of the primes from the wheel of 210: the residue classes coprime to $2 \cdot 3 \cdot 5 \cdot 7$, the race among them, the conductor lattice of their characters, the explicit formula tying the primes to the zeros of the Riemann zeta function, and the three walls at which honest exposition stops — the Riemann Hypothesis, the twin primes, and the hardness of factoring. This book assumes that one. It does not re-derive the wheel or re-introduce the explicit formula; it takes the headline ideas of the first book — including the Hilbert–Pólya dream, the Berry–Keating model, and Connes's trace formula, all met there in outline — and goes one level deeper into the single most seductive face of the subject: the idea that the zeros are the *spectrum* of something, and that to understand the primes one should find the instrument whose notes they are.

The stance is the same as the first book's, and it is worth stating plainly. Everything credited to the established literature is the work of the people named; the metaphors are marked as metaphors and the mechanisms as mechanisms; and where a calculation is the book's own, it is presented with the numbers and the means to reproduce it. The book does not pass the three walls, and it does not pretend to. It ends at them on purpose, because the most useful thing a tour can do at a wall is show exactly where it stands and why it is hard, rather than tell a story that walks through it.

A few conventions carry over. The word *subprime* means, throughout, a composite number coprime to 210 whose smallest prime factor exceeds 7. Each chapter opens with a short SCOPE and closes with a section called *What this is and is not*, separating the established from the merely measured from the frankly conjectural; where a chapter has a figure, a closing note prints the short program that produced it, and the longer computations are reproduced by the verification programs that accompany this book. The voice is plain and declarative on purpose; the subject supplies enough difficulty without ornament.

The book is in four parts. *The statistics* looks at the zeros as a population and asks what kind of spectrum they resemble. *The operator dream* takes the Hilbert–Pólya idea seriously enough to try to build the operator, and watches the attempts fail in instructive ways. *Hearing and reading* asks what the primes can tell us about the zeros — what can be detected, what cannot be resolved, and which apparent laws do not survive being pushed. *The walls* walks up to the three obstacles again, through the spectral lens, and finds them standing where the first book left them.

Acknowledgements

The author gratefully acknowledges the use of Anthropic's Claude as a computational and drafting assistant during the preparation of this work. Any mathematical results or interpretations presented here remain entirely the responsibility of the author. — J.M.

Contents

Part I

The statistics

Chapter 1

The Symmetry of a Family

Book One looked at the zeros of one zeta function among themselves and found the statistics of the eigenvalues of a large random Hermitian matrix. This chapter makes the move Book One did not: from a single spectrum to a whole family of them. The governing fact, due to Katz and Sarnak, is that the local statistics deep in the spectrum are the same for every family — the sine-kernel hole already met in Book One — but each family also carries a symmetry type, *unitary or orthogonal or symplectic, and that type is written in one place only: the way the very lowest zeros sit against the central point. A symplectic family pushes its lowest zeros away from the centre; an orthogonal family pulls them in; a unitary family is indifferent. And there is exactly one arithmetic world in which all of this is a theorem rather than a hope — curves over a finite field, where the zeros really are the eigenvalues of a real operator and the Riemann Hypothesis is Weil's theorem. Over the rational numbers the same statistics hold to every digit anyone has computed, and stay conjectural. This chapter is expository.*

1.1 Scope

This chapter is expository and reports established mathematics. It assumes the previous book's spectral chapter — unfolding, level repulsion, Montgomery's pair correlation, and the Hilbert–Pólya idea — and goes one step past where that chapter stopped. There, the object was the single sequence of zeros of ζ. Here the object is a *family* of L-functions, and the new content is threefold: that the members of a family are statistically independent of one another, that families sort into three symmetry types according to the behaviour of their lowest zeros, and that this sorting is a genuine theorem in one setting and a well-supported conjecture in ours. The chapter introduces the independence of distinct L-functions seen on the wheel's seven, the Katz–Sarnak symmetry types, the one-level density that detects them, the function-field world where they are proved, and the precise sense in which the rational world falls short. It proves nothing new. The one thing this circle of ideas would most like to hand over — an operator over the rationals to play the role the Frobenius operator plays over a finite field — is exactly what the next chapter is unable to find, and Section 1.7 is careful about that line.

1.2 From one spectrum to a family

A single zeta function gives you a single spectrum: one critical line, one ladder of zeros climbing it. Book One showed that ladder, once unfolded to unit spacing, keeps the statistics of a random Hermitian matrix. But a statistical law wants more than one sample. The eigenvalues of *one* large matrix already display the law; the law itself, though, is a statement about the *ensemble* — about what happens on average as the matrix is drawn again and again. The arithmetic analogue of drawing the matrix again is changing the L-function.

So the natural object is not ζ alone but a family of L-functions that resemble it: the Dirichlet L-functions of a fixed modulus; the quadratic L-functions $L(s, \chi_d)$ as d runs over fundamental discriminants; the L-functions of the quadratic twists of a fixed elliptic curve. Each member has its own critical line and its own lowest zeros. Averaging a statistic across the family is the arithmetic version of averaging over the ensemble of matrices. The seven real characters of the wheel are the smallest such family sitting in plain sight — seven quadratic L-functions, one per spoke-symmetry of Book One. Seven is far too few to see anything statistical, a point we return to; but they fix the idea, and they will turn out to belong to a definite one of the three types.

1.3 The bulk is one law for everyone

The first thing to say about families is a negative: deep inside the spectrum, they are all the same. The two-point statistic of the high zeros — Montgomery's pair correlation from Book One,

$$R_2(u) = 1 - \left(\frac{\sin \pi u}{\pi u}\right)^2, \tag{1.1}$$

with its deep hole at small separation — is the *same* function for every family, and it is the same as the bulk two-point law of the unitary, orthogonal, and symplectic random-matrix ensembles alike. This is the content of *universality*: the local spacing of the high zeros does not know which family it came from. Panel A of Figure 1.1 is this one law. If the bulk were all we could measure, the three symmetry types would be indistinguishable.

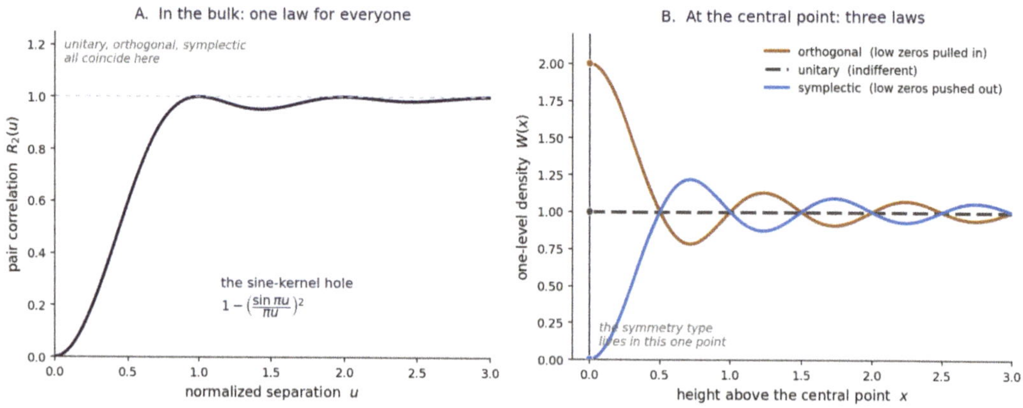

Figure 1.1: Where a family hides its symmetry. **A:** the bulk two-point statistic, Eq. (1.1), is the same sine-kernel hole for every L-function in every family, and the same for the unitary, orthogonal, and symplectic ensembles — the bulk cannot tell them apart. **B:** the one-level density $W(x)$, which measures how the lowest zeros sit at height x above the central point. Here the three types separate, and only at the central point itself: a symplectic family has $W(0) = 0$ (lowest zeros repelled), a unitary family $W(0) = 1$ (indifferent), an orthogonal family $W(0) = 2$ (lowest zeros attracted). The symmetry type lives in that single point.

1.4 The families are independent draws

Universality says every family obeys the same bulk law. It does not say the families have anything to do with one another, and they do not. Take the seven real L-functions of the wheel and measure, for each one alone, the average ratio of consecutive gaps between its zeros — a single number, needing no unfolding, that a unitary spectrum drives to 0.603 and an uncorrelated one to 0.386. Each of the seven returns about 0.62 — at the unitary end, a shade above the asymptotic 0.603 and far from 0.386: each spectrum, on its own, repels exactly as Eq. (1.1) demands.

Now pour all seven sets of zeros into one list and measure again. The number collapses to about 0.39 — the value for points with no correlation (Figure 1.2). The repulsion each family carried alone is gone from the merged sequence: its gaps are those of a structureless list, the sine-kernel hole of Eq. (1.1) filled flat. Rigid spectra can superimpose into a structureless one only when they are mutually independent: any dependence among the seven — coincident zeros, correlated ordinates, or a rigidity shared across the families — would show as a departure from the Poisson value, in one direction or the other, and none appears. This is the statistical independence of distinct L-functions that Rudnick and Sarnak established, seen directly on the wheel's seven. Like the bulk law, it is a statement about how the zeros are distributed and not about where they lie, and it makes no claim about the Riemann Hypothesis.

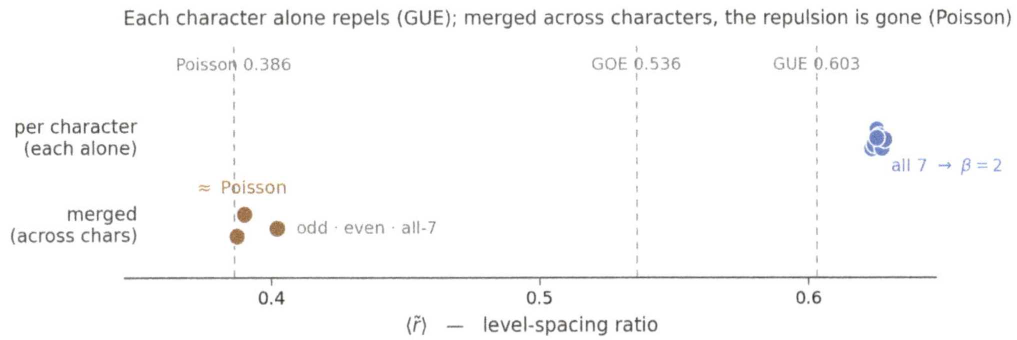

Figure 1.2: Independence by collapse. For each of the seven real L-functions of the wheel, measured alone, the level-spacing ratio sits at the unitary end (blue, ≈ 0.62, just above the $\beta = 2$ value 0.603): the zeros repel, as in Eq. (1.1). Pooling the seven — as the odd four, the even three, and all seven — sends the ratio to the Poisson value (orange, the three points near 0.39): the repulsion is gone. Rigid spectra merge to a structureless one only when they are independent, so the collapse is the independence of the families made visible. Reference values: Poisson 0.386, GOE 0.536, GUE 0.603.

1.5 The symmetry lives at the central point

What separates the families is not how their zeros sit among themselves high up, but how the *lowest* zeros sit relative to one distinguished height: the central point $s = \frac{1}{2}$, where the functional equation pivots. The statistic that sees this is the *one-level density* — loosely, the averaged density of low zeros at height x above the centre, measured across the family. Its scaling form is one of three closed expressions, shown in Panel B:

$$W_{\text{unitary}}(x) = 1, \tag{1.2}$$

$$W_{\text{symplectic}}(x) = 1 - \frac{\sin 2\pi x}{2\pi x}, \tag{1.3}$$

$$W_{\text{orthogonal}}(x) = 1 + \frac{\sin 2\pi x}{2\pi x}. \tag{1.4}$$

At the centre, $x \to 0$, the kernel $\sin(2\pi x)/(2\pi x) \to 1$, so the three densities take the values 0, 1, and 2. That is the whole distinction, and it is a statement about a single point. A *symplectic* family clears a gap at the centre — its lowest zeros are pushed up, away from $s = \frac{1}{2}$. An *orthogonal* family does the opposite, crowding its lowest zeros down toward the centre. (The density $W_{\text{orthogonal}}$ of Eq. (1.4) is the even orthogonal SO(even) density; the odd orthogonal SO(odd) case instead carries a forced zero at the centre.) A *unitary* family does neither; its low zeros are as indifferent to the centre as to any other height.

Which family is which is decided by the arithmetic symmetry each member carries. The full set of Dirichlet characters to a modulus is a *unitary* family. The quadratic characters — the real ones, the kind the wheel is built from — form a *symplectic* family. The quadratic twists of an elliptic curve are *orthogonal*, and split further by the sign of the functional equation into an even part and an odd part. So the wheel's seven real characters are, in principle, members of a symplectic family: their lowest zeros should sit a touch higher than a unitary family's would. The qualification "in principle" is doing real work. To *see* the curve of Eq. (1.3) you need the lowest zeros of thousands of quadratic L-functions, smoothed into a density. Seven characters give you seven numbers and no curve at all. (The independence of the previous section escaped this shortage: it pooled the *zeros* of the seven, which run to the thousands, while a symmetry type is a statistic across *family members* — and of those, seven is far too few.) This is the first, quiet appearance of a problem that will dominate the second half of the book: the law is real, and the data in hand is too thin to resolve it.

1.6 The one world where it is a theorem

Everything so far is conjecture over the rational numbers. There is a different arithmetic world where the same statements are theorems, and the difference between the two worlds is the heart of this book.

Replace the rational numbers by the functions on a curve over a finite field $\mathbb{F}_q$. The L-functions attached to such a curve are not mysterious infinite products that might or might not have their zeros on a line; they are *polynomials*. By Weil, and in full generality by Deligne in 1974, the reciprocal roots of these polynomials are $q^{1/2}$ times numbers of absolute value one — the eigenvalues of the Frobenius operator acting on the cohomology of the curve. In this world the Riemann Hypothesis is not open. It is Weil's theorem. The zeros lie on the critical line not because a real operator has a real spectrum — Frobenius is not self-adjoint — but because of *purity*: the Weil–Deligne weight argument forces the Frobenius eigenvalues to absolute value $q^{1/2}$, which is exactly the Riemann Hypothesis for the curve. An operator whose eigenvalues are the zeros genuinely exists here, realizing the Hilbert–Pólya dream over a finite field rather than over $\mathbb{Q}$; what places the zeros on the line, though, is purity, not self-adjointness.

Katz and Sarnak then proved, in 1999, that as the field grows the zeros of families of these L-functions follow exactly the symmetry-type laws of Panel B. And the symmetry type is not an analogy there: it is the *monodromy group* of the family, the group through which the Frobenius eigenvalues turn as the curve varies. "Unitary," "orthogonal," "symplectic" name actual structure groups, and Deligne's equidistribution theorem makes the random-matrix average an equality rather than a hope. Over function fields, in short, the spectrum is an operator's spectrum, the symmetry is a group, and the statistics are theorems.

Back over $\mathbb{Q}$, the best available results are real but bounded. Özlük and Snyder proved in 1999 (later extended by Rubinstein) that the quadratic family has the symplectic one-level density of Eq. (1.3) — but only when tested against functions whose Fourier transform is supported in a limited window. That is the very same restriction Montgomery met in Book One, where the pair correlation could be confirmed only for test functions of limited support. Inside the window it is a theorem; outside it, the symmetry type is conjecture backed by overwhelming numerics.

1.7 What this is and is not

What is established is of two kinds. The bulk universality — the sine-kernel hole of Eq. (1.1), shared by every family — is on the firmest footing the subject offers. And over function fields the entire symmetry-type theory is a theorem, because there the zeros are the eigenvalues of the Frobenius operator and the Riemann Hypothesis is Weil's. Over the rationals, the symplectic one-level density of the quadratic family is a theorem for test functions of restricted support, by Özlük–Snyder (and Rubinstein).

What is not established is the rest of it over $\mathbb{Q}$: the symmetry types beyond the restricted window, for every family, remain conjectural. The gap is sharp, and it is worth naming exactly, because the rest of the book lives in it. The function-field case is a theorem and the rational case is not, and the difference between them is not a difference of evidence — the numerical agreement over $\mathbb{Q}$ is as good as numerical agreement ever gets. It is a difference in the *kind of object* available. Over a finite field there is an operator, Frobenius on cohomology, whose eigenvalues are the zeros; the symmetry types fall out of its monodromy as a matter of proof. Over $\mathbb{Q}$ there is no such operator, and so the same statistics can only be observed, not derived.

Book One said the zeros of ζ behave *as though* they were a spectrum. This chapter sharpens that into a contrast. In one arithmetic world the zeros simply *are* a spectrum, and one can watch the symmetry types emerge from the operator. In ours, we have the identical statistics and no operator to hang them on. Whether such an operator exists over $\mathbb{Q}$ — and what would have to be true of it — is the subject of the next chapter.

Reproducing the figures

```
import numpy as np

# Panel A: the bulk two-point law, shared by all symmetry types
def sinc1(u):                   # sin(pi u)/(pi u), = 1 at u = 0
    s = np.ones_like(u, dtype=float); nz = u != 0
    s[nz] = np.sin(np.pi*u[nz])/(np.pi*u[nz]); return s
u  = np.linspace(1e-4, 3, 800)
R2 = 1 - sinc1(u)**2            # Montgomery pair correlation

# Panel B: one-level densities; the kernel is sin(2 pi x)/(2 pi x)
def sinc2(x):                   # = 1 at x = 0
    s = np.ones_like(x, dtype=float); nz = x != 0
    s[nz] = np.sin(2*np.pi*x[nz])/(2*np.pi*x[nz]); return s
x      = np.linspace(0, 3, 800)
W_unit = np.ones_like(x)            # unitary       : W(0) = 1
W_symp = 1 - sinc2(x)               # symplectic    : W(0) = 0
W_orth = 1 + sinc2(x)               # orthogonal    : W(0) = 2

# central-point values that carry the whole distinction
print(1 - sinc2(np.array([0.0])))   # [0.] symplectic: lowest zeros repelled
print(1 + sinc2(np.array([0.0])))   # [2.] orthogonal: lowest zeros attracted

import numpy as np
```

```
# Independence figure: the level-spacing ratio r~ of a sorted ordinate list.
#   no unfolding needed; ~0.603 for GUE (beta=2), ~0.386 for uncorrelated points.
def rtilde(g):
    g = np.sort(np.asarray(g, float)); s = np.diff(g)
    r = np.minimum(s[1:], s[:-1]) / np.maximum(s[1:], s[:-1])
    return float(r.mean())

# each of the seven real L-functions, alone, gives ~0.62 (GUE);
# the seven ordinate lists poured into one give ~0.39 (Poisson) -- the collapse.
# the per-character and merged ordinates are produced by the verification
# programs accompanying this book.
```

Part II

The operator dream

Chapter 2

What Would Count as an Operator

The previous chapter ended on a single missing object: over a finite field the zeros are the eigenvalues of the Frobenius operator and the Riemann Hypothesis is a theorem, while over the rationals we have the same statistics and no operator to hang them on. This chapter states exactly what such an operator would have to be, and then watches the two most natural attempts to build one fail — in opposite ways. A candidate must pass a two-part test: it must be defined without already knowing the zeros, *and its eigenvalues must* be *the zeros. The first attempt builds an object whose spectrum is exactly the zeros, by writing the zeros into it by hand; it has the right spectrum and fails the first half of the test. The second builds the most natural operator the primes offer, intrinsic and honest, and finds that its spectrum is the logarithms of the integers, not the zeros; it passes the first half and fails the second. The two failures are not bad luck. They are the two walls of a corridor, and the corridor is narrow.*

2.1 Scope

This chapter is partly expository and partly a pair of worked computations. It assumes the previous book's primon-gas chapter, whose bosonic gas of prime "modes" returns here as a genuine operator, and it assumes the Hilbert–Pólya idea as that book stated it. The new content is a sharpening and two examples. The sharpening is the two-part test that any Hilbert–Pólya operator must pass. The examples are two operators built explicitly and examined honestly: one with the zeros as its spectrum and no independent definition, one with an impeccable independent definition and the wrong spectrum. Both computations are exact and checkable; neither produces a Hilbert–Pólya operator, and Section 2.6 says why in the sharpest terms the chapter can manage. The genuinely promising candidate — Berry and Keating's — is held over to the next chapter.

2.2 The dream, stated exactly

The Hilbert–Pólya idea, as Book One gave it, is that the ordinates γ_n of the nontrivial zeros are the eigenvalues of some self-adjoint operator H. If that were so, the γ_n would be real because the spectrum of a self-adjoint operator is real, and the zeros $\frac{1}{2} + i\gamma_n$ would be pinned to the critical line not by a delicate analytic balance but by the reality of a quantum spectrum. The Riemann Hypothesis would follow at a stroke (Figure 2.1).

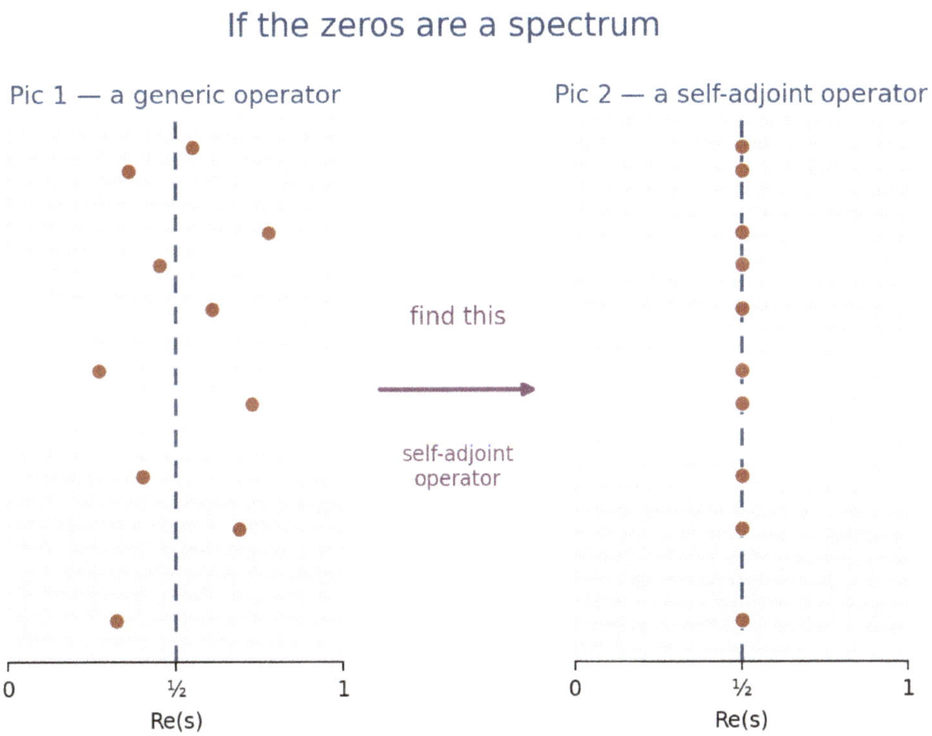

Figure 2.1: The shape of the dream. Were the zero ordinates γ_n the spectrum of a self-adjoint operator, that spectrum would be real, and every zero would lie on the line $\mathrm{Re}(s) = \frac{1}{2}$ (Pic 2). Pic 1 shows what a *generic* operator would permit; the arrow is the unsolved part—no such operator has been constructed. Schematic.

To turn this from a wish into a target, we ask what would actually count as a solution. Two conditions, and a candidate must meet both.

(I) **Independence.** The operator H must be defined without already knowing the zeros. It should be built from arithmetic — from the primes, from a space of functions, from a geometry — by a recipe that never mentions the γ_n. An operator you can only write down after someone hands you the zeros explains nothing, because the explaining was done by whoever found the zeros.

(II) **Correct spectrum.** The eigenvalues of H must be the γ_n themselves — not a sequence that resembles them, not a sequence with the same statistics, but the actual ordinates.

The two conditions sound almost the same on first reading, and they are almost opposite in practice. Condition (II) alone is easy; condition (I) alone is easy; the difficulty is that no known construction meets both at once. The one place they have been met together is the function-field world of the previous chapter, where the Frobenius operator is defined intrinsically from the geometry of a curve — it never mentions the zeros (I) — and its eigenvalues *are* the zeros of the curve's L-function (II). That is the whole reason the Riemann Hypothesis is a theorem over a finite field and an open problem over the rationals. With the bar set, we try to clear it.

2.3 The first failure: a spectrum made to order

Take the zeros as given — the first several hundred ordinates γ_n, computed to high precision — and ask for an object whose spectrum is exactly those numbers. There is an embarrassingly direct way to get one. Form the function

$$K(\xi) \;=\; \sum_n \cos(\gamma_n \xi), \tag{2.1}$$

a finite sum of cosines, one per ordinate. Its frequency content — the thing a physicist would call its spectrum — is a set of spikes sitting exactly at the γ_n, because that is what each cosine contributes and nothing else is present. Panel A of Figure 2.2 is this spectrum: the zeros, reproduced perfectly.

One can do this with more care and get more structure. Pair each ordinate with the midpoint $m_n = \frac{1}{2}(\gamma_n + \gamma_{n+1})$ to its successor and form the windowed difference

$$K(\xi) \;=\; \frac{1}{|I|}\sum_{n\in I}\big[\cos(\gamma_n \xi) - \cos(m_n \xi)\big], \tag{2.2}$$

a trigonometric polynomial of some hundreds of cosine terms — in one natural windowing, equally many at the ordinates and at the midpoints. This object has a genuine closed form, a genuine asymptotic behaviour, and even a genuine combinatorial topology: which zeros pair with which midpoints depends on the arithmetic of the underlying wheel, so the kernel carries real structure that took real work to describe. None of that rescues it. Its spectrum is a signed sum of spikes at the ordinates and the midpoints, and the ordinates are in the spectrum for one reason only: we wrote them in. The object satisfies condition (II) and violates condition (I) completely. It is not a window onto the zeros; it is a photograph of them, and a photograph predicts nothing about its subject.

This is the first lesson of the operator dream, and it is worth stating bluntly because the literature is full of subtler versions of the same move: *having the right spectrum is no achievement if the spectrum was the input.* Structure is not the issue — this kernel has plenty of structure. Independence is the issue.

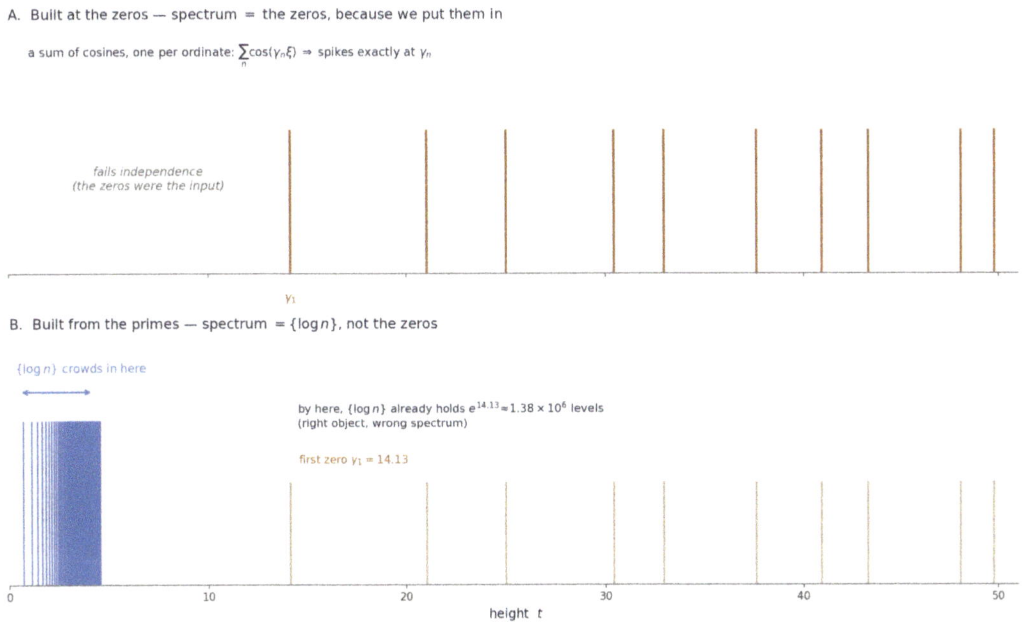

Figure 2.2: Two natural operators, two ways to miss. **A:** an object built as a sum of cosines at the ordinates, Eq. (2.1). Its spectrum is the zeros exactly — because the zeros were written into it. It has the right spectrum and no independent definition. **B:** the primon-gas Hamiltonian of Book One, built only from the primes. Its spectrum is $\{\log n\}$, the logarithms of the integers, which thin out near the origin and crowd together at the top of the range, never taking the shape of the zeros (faint, for reference, at 14.13 and beyond). By the height of the first zero, this spectrum already holds about 1.38 million levels. It has an impeccable independent definition and the wrong spectrum.

2.4 The second failure: the right object, the wrong spectrum

Now do the honest thing and refuse to look at the zeros at all. Build an operator from the primes alone. The natural one is the Hamiltonian of Book One's primon gas: a bosonic Fock space with one mode for each prime, and

$$\widehat{H} = \sum_p (\log p)\, \hat{n}_p, \tag{2.3}$$

where $\hat{n}_p$ counts the quanta in the mode of the prime p and each such quantum costs energy $\log p$. This is as intrinsic as an operator can be. It mentions no zeros; it is written down from the primes and the single rule that a prime p carries energy $\log p$. It passes condition (I) without an argument.

Its spectrum is immediate and beautiful. A state with k_p quanta in each prime mode has energy $\sum_p k_p \log p = \log\bigl(\prod_p p^{k_p}\bigr) = \log n$, where n is the integer with that factorization. By unique factorization every positive integer arises exactly once, so the spectrum is

$$\operatorname{spec} \widehat{H} = \{\log n : n = 1, 2, 3, \dots\}, \tag{2.4}$$

the logarithms of the integers, each appearing once. This is the object whose partition function $\sum_n e^{-s \log n} = \sum_n n^{-s}$ is the zeta function itself — which is the sharp irony of the construction. The zeta function is the primon gas's partition function, and the primon gas's spectrum is $\{\log n\}$. Zeta enters through its Dirichlet series, not through its zeros, and the operator never sees a single γ_n.

The mismatch is total, and Panel B makes it visible. The logarithms of the integers thin out near the origin and crowd together at the top of the range — $\log 2 \approx 0.69$, $\log 3 \approx 1.10$, and the gaps shrink like $1/n$, so the spectrum becomes a near-continuum — while the zeros begin only at $\gamma_1 = 14.13$ and stay discrete, repelling one another. The cleanest way to see how far apart these two sequences are is to count. The number of integers with $\log n \le t$ is $\lfloor e^t \rfloor$, growing exponentially in t. The number of zeros with $\gamma_n \le t$ grows like $\frac{t}{2\pi} \log \frac{t}{2\pi}$, barely faster than linearly. By the height of the very first zero, $t = 14.13$, the primon spectrum already contains $e^{14.13} \approx 1.38$ million levels, and the zeros contain one. No reweighting, no change of variable, repairs a discrepancy of that order. The operator is the right kind of thing and has entirely the wrong spectrum. It passes (I) and fails (II).

A word on names, because they cause real confusion. The Hamiltonian Eq. (2.3) is sometimes called a Berry–Keating Hamiltonian. It is not the one Berry and Keating proposed. Theirs is a different operator, built on a single classical degree of freedom rather than a gas of prime modes, and it is the subject of the next chapter. Equation (2.3) is the free primon gas of Book One, and its spectrum is $\{\log n\}$. Keeping the two apart is the difference between an operator that fails the second test outright and one that might, conceivably, pass it.

2.5 Why the two failures are one

The two operators fail opposite halves of the test, and that opposition is the real content of the chapter. Push on either one and you are pushed toward the other's failure. Try to repair the cosine kernel of Eq. (2.1) by giving it an intrinsic definition — some rule that produces those cosines without being handed the ordinates — and you are asking for exactly the operator nobody has: condition (I) is the whole problem. Try to repair the primon gas of Eq. (2.3) by bending its spectrum from $\{\log n\}$ toward the zeros, and every natural bending — a different weight on each prime, a chemical potential, an interaction between modes — moves the spectrum to some other arithmetic sequence built from the primes, never to the zeros, because the zeros are not a simple functional of the primes in that way; they are the primes' Fourier-dual, the subject of Book One's diffraction chapter, and Fourier duals do not appear as the energy levels of the thing they are dual to.

So the corridor is narrow on purpose. An intrinsic construction from the primes wants to give you $\{\log n\}$ or a relative of it; a construction with the zeros as spectrum wants the zeros as input. The function-field world escapes the corridor because there the dual variable is itself geometric — Frobenius acts on a cohomology that the zeros live in — so intrinsic definition and correct spectrum coincide. Over the rationals no one has found the room where they coincide. That room, if it exists, is what the next two chapters go looking for.

2.6 What this is and is not

What is established here is two exact facts. The cosine kernel of Eqs. (2.1) and (2.2) has spectrum equal to the ordinates by construction; this is a triviality dressed as a computation, and naming it as such is the point. The primon-gas Hamiltonian of Eq. (2.3) has spectrum $\{\log n\}$; this is a clean theorem about a clean operator, proved in one line from unique factorization. Both are mathematics, and neither is in doubt.

What is not established — by anyone — is a Hilbert–Pólya operator over the rationals. The two constructions here are not failed attempts at one; they are demonstrations of the two ways the attempt characteristically fails. And it is important to be honest about the status of the test itself. There is no theorem that says no such operator exists. The two-part test is not an impossibility proof; it is a statement of the bar, and a map of the terrain just below it. Many serious programs — Berry and Keating's, Connes', de Branges' — are attempts to find the narrow room where conditions (I) and (II) hold together, and the chapters that follow take each at its true status: real ideas, real partial results, no operator yet. What this chapter shows is only this. The easy ways to get the zeros as a spectrum cheat, and the honest operator the primes hand you has the logarithms of the integers for a spectrum, which is the wrong sequence by a factor of a million levels before the first zero is reached.

Reproducing the figure

```
import numpy as np
from mpmath import zetazero
g = np.sort(np.array([float(zetazero(k).imag) for k in range(1, 31)]))

# Panel A: spectrum of the cosine kernel sum_n cos(gamma_n xi)
#   = spikes at the ordinates themselves (they were the input)
zeros_in_window = g[g <= 52]                  # gamma_1 .. ~gamma_10

# Panel B: spectrum of the primon-gas Hamiltonian  H = sum_p (log p) n_p
#   eigenvalue of an occupation state {k_p} is sum k_p log p = log n
#   so spec(H) = { log n : n = 1, 2, 3, ... }
n    = np.arange(2, 60)
logn = np.log(n)                              # all land in [0.69, ~4.07]

# how far apart the two spectra are: count levels below height t
#   primon:  #{ log n <= t } = floor(e**t)
#   zeros :  #{ gamma_n <= t } ~ (t/2pi) log(t/2pi)
import math
print(math.floor(math.exp(g[0])))            # 1376046  levels below gamma_1 = 14.13
```

Chapter 3

Primes as Orbits

The last chapter left a corridor with two walls: an intrinsic operator from the primes gives the logarithms of the integers, and an operator with the zeros as spectrum has to be handed the zeros. This chapter walks the one candidate that gets furthest down the corridor. Berry and Keating's classical Hamiltonian $H = xp$ is built from nothing but a position and a momentum — it never mentions the zeros, so it passes the independence test cleanly — and its regularized semiclassical level count reproduces the average density of the Riemann zeros exactly. *That is the deepest partial success in the subject. But the average is all it gives. The individual zeros live in the fluctuation around the average, and the fluctuation is supplied not by xp but by the primes, entering the explicit formula in precisely the shape of a sum over periodic orbits — one orbit per prime, of period* $\log p$. *The explicit formula, read this way, is a trace formula. The catch is that we have the trace formula and the orbits but not the dynamical system they belong to, and the reason that system is so hard to find is the same reason its spectrum looks like the zeros in the first place: it would have to be chaotic.*

3.1 Scope

This chapter is expository, building on two chapters of Book One — the diffraction chapter that read the zeros as the Fourier-dual of the primes, and the primon-gas chapter — and on the two-part test of the previous chapter. The new content is the Berry–Keating proposal at its true weight: what $H = xp$ achieves, which is real and exact, and what it does not, which is everything that distinguishes one zero from the next. The chapter then rereads Book One's explicit formula as a trace formula in the sense of quantum chaos, with the primes as periodic orbits, and explains why that reading is a genuine structural match and still not a proof. It closes on the reason the whole programme is hard, which is the reason it is attractive: the spectrum behaves like that of a chaotic system, and chaos resists exactly the kind of forward computation a proof would need. Connes' and de Branges' programmes, which try to supply the missing system by other means, are named here and taken up later.

3.2 An operator that passes the first test

Berry and Keating proposed, as the classical mechanics underneath the Riemann zeros, the simplest Hamiltonian that is not a harmonic oscillator:

$$H = x\,p, \tag{3.1}$$

a single particle on a line, with energy equal to position times momentum. The trajectories are hyperbolae $xp = E$ in the phase plane. Nothing in this definition knows anything about ζ; it is written from a position and a momentum and a product. So $H = xp$ passes the independence test of the previous chapter without discussion — and it is the first candidate in this book to do so while also getting something quantitative right.

Here is what it gets right. To count the quantum levels of a system below energy E, the semiclassical rule is to measure the area of the phase-space region with $H \leq E$ and divide by 2π (in units where Planck's constant is one). For $H = xp$ that region is unbounded — a hyperbola encloses infinite area — so one must regularize, by refusing to resolve phase-space cells smaller than a single quantum: $|x| \geq \ell_x$, $|p| \geq \ell_p$. With that cutoff the area below E is finite, and the count of levels below height T comes out as

$$N_{\text{smooth}}(T) \;=\; \frac{T}{2\pi}\log\frac{T}{2\pi} - \frac{T}{2\pi} + \frac{7}{8} + \cdots \tag{3.2}$$

This is not merely the right order of magnitude. It is, term for term, the smooth part of the Riemann–von Mangoldt counting function for the zeros of ζ — the same $N_{\text{smooth}}(T)$ Book One used to unfold them, down to the constant $\frac{7}{8}$. Panel A of Figure 3.1 shows the curve of Eq. (3.2) laid over the actual staircase of zeros: at $T = 100$ the formula gives 29.00 and there are exactly 29 zeros below that height. A Hamiltonian written from a position and a momentum, regularized by a single quantum cell, counts the Riemann zeros on average to the last term. That is a real and surprising fact, and it is the reason xp is taken seriously.

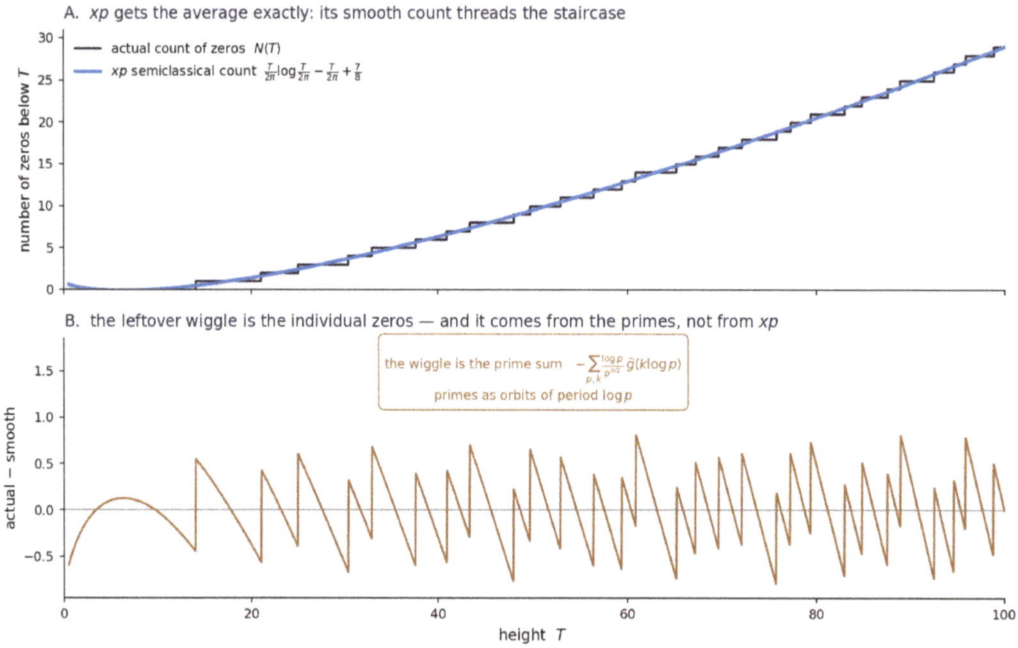

Figure 3.1: What xp gives and what it misses. **A:** the regularized semiclassical count of $H = xp$, Eq. (3.2) (smooth curve), threads the staircase of actual zeros (steps); it reproduces the average density of zeros term for term. **B:** the leftover, actual count minus smooth count. This fluctuation — a unit jump at each zero, sloping between — is where the individual zeros live, and xp does not produce it. In the explicit formula it is exactly the sum over primes, each prime acting as a periodic orbit of period $\log p$.

3.3 The average and the fluctuation

Now the limitation, which is visible the moment one subtracts. Panel B is the difference between the actual count of zeros and the smooth count of Eq. (3.2): a sawtooth that jumps up by one at each zero and slopes gently down between them. The smooth curve is the *average* behaviour; the sawtooth is the *fluctuation*, and every fact that distinguishes one zero from another — where exactly γ_1 sits, how far it is to γ_2 — is contained in the fluctuation, not in the average.

$H = xp$ produces the average and nothing else. It is a free system, in the sense that its regularized phase space has no internal structure beyond the cutoff; its level density is smooth by construction. The fluctuation has to come from somewhere else, and in the Riemann problem there is only one other place it can come from: the primes. This is the content of Book One's diffraction chapter restated as a budget. The zeros are an average part, set by a smooth count that xp reproduces, plus a fluctuating part, set by the primes. xp owns the first line of the budget and contributes nothing to the second.

3.4 The explicit formula as a trace formula

The way the primes supply the fluctuation has a shape, and the shape is the point of the chapter. Book One's explicit formula relates a sum over the zeros to a sum over the primes; in one standard form, for a suitable test function g and its transform $\widehat{g}$,

$$\sum_n g(\gamma_n) = \underbrace{(\text{smooth archimedean term})}_{\text{the average}} - \underbrace{\sum_p \sum_{k\geq 1} \frac{\log p}{p^{k/2}}\, \widehat{g}(k \log p)}_{\text{the fluctuation}} . \tag{3.3}$$

Read the two sides in the language of quantum chaos. On the left is a sum over the *energy levels* of a system — the zeros. On the right, the first term is the smooth density of states, the very term xp supplies. The second term is a sum over the primes, and each prime contributes a wave $\widehat{g}(k \log p)$ of "length" $k \log p$, weighted by $(\log p)\, p^{-k/2}$. That is exactly the form of Gutzwiller's trace formula, the central identity of quantum chaos, in which the fluctuating density of states of a chaotic system is a sum over its classical *periodic orbits*: each orbit contributes a wave whose phase is its length and whose amplitude is set by how unstable it is. The dictionary is clean. A primitive periodic orbit is a prime p; its period is $\log p$; traversing it k times gives the repetition $k \log p$; its amplitude $(\log p)\, p^{-k/2}$ is the period times a stability factor $p^{-k/2}$. The explicit formula, in other words, *is* a trace formula — and unlike almost everything else in this book, it is a theorem, stated by Riemann, first proved rigorously by von Mangoldt, and cast in general form by Weil, not a hope.

The primon gas of Book One makes the period $\log p$ concrete rather than metaphorical. Evolve that gas in time: the mode of the prime p rotates by the phase $p^{it} = e^{i(\log p)t}$, so in the gas each prime is literally an oscillator of frequency $\log p$. The "orbit of period $\log p$" is not an analogy borrowed from billiards; it is the natural motion of the prime mode that Book One already built. The primes are oscillators, the zeros are the spectrum that their interference would produce, and Eq. (3.3) is the bookkeeping that ties the two together.

3.5 Why it is a near-miss and not a proof

Everything above is real, and none of it is a proof of anything. The reason is worth stating precisely, because xp and the trace formula are the place where the operator dream comes closest, and seeing exactly how close is seeing exactly where the wall is.

In an honest application of Gutzwiller's formula one starts with a dynamical system, finds its periodic orbits and their periods and stabilities *from the dynamics*, and the trace formula then follows as a theorem about that system. The Riemann case runs the logic backwards. We have the trace formula, Eq. (3.3), as a theorem. We can read off the orbits and their periods — they are the primes, with periods $\log p$. What is missing is the system: no one has exhibited a dynamical system, classical or quantum, whose periodic orbits are the primes with these periods and these amplitudes. The Hilbert–Pólya operator would *be* that system, so "find the system whose orbits are the primes" and "find the operator whose spectrum is the zeros" are the same unsolved problem wearing two costumes. The trace formula is a true identity between two sums you would like to compute; it is not a machine that takes the primes in and hands the zeros out.

Three specific features show that the missing system, if it exists, is not an ordinary one. First, the periodic-orbit sum in Eq. (3.3) carries an overall *minus* sign: the primes enter as subtractions, so the zeros behave less like the emission lines of a spectrum and more like absorption lines — gaps where energy is missing. (It was Connes who turned this sign into a precise spectral statement, the zeros as an absorption spectrum on a large symmetry space; we return to it.) Second, xp is integrable rather than chaotic (and in fact breaks time-reversal symmetry), so although its phase-space volume reproduces the average counting function, it cannot generate the chaotic fluctuations the individual zeros require. Third, xp without its regularization has a continuous spectrum — the whole real line — not the discrete ladder of the γ_n; the discreteness has to be installed by a boundary condition or a cutoff, and every known way of installing it either fails to give the zeros or smuggles in a choice that amounts to knowing them. On the two-part test, then, xp lands in a third position, different from either failure of the previous chapter: it passes independence, and it passes the *average* half of the correct-spectrum condition exactly, while missing the fluctuating half that contains the actual zeros. It is the most honest near-miss the subject has.

3.6 Why chaos is the wall

There is a temptation to read all of this as a near-miss that one more clever idea will close. The last thing this chapter wants to leave is the reason that is unlikely, and it is the same fact that makes the trace-formula picture beautiful: the system, if it exists, is chaotic.

A chaotic system is precisely one whose long trajectories are unpredictable and whose periodic orbits proliferate exponentially, and the trace formula for such a system is a delicate object. Its periodic-orbit sum does not converge in the ordinary sense; it must be handled by analytic continuation, by careful resummation, by using the very spectrum it is meant to produce to make sense of itself. This is visible already in Eq. (3.3): the prime sum on the right converges only conditionally, and giving it a value requires information equivalent to knowing where the zeros are. The identity is genuinely two-directional — it constrains primes by zeros as much as zeros by primes — which is why it cannot be run as a one-way computation. You cannot feed in the primes and grind out the zeros, because the grinding only makes sense once the zeros are in hand.

So chaos sits on both sides of the ledger. It is the best available explanation for why the zeros look the way they do: a chaotic spectrum with broken time-reversal symmetry would have exactly the level repulsion and pair correlation Book One measured. And it is the reason the zeros cannot be computed from below: a chaotic system has no closed form for its individual levels, its trace formula is conditionally convergent and self-referential, and the dynamical system that would anchor the whole picture has never been found. The same word explains the statistics and forbids the shortcut.

That is what it means for this to be a wall and not a gap, and it is why the chapters that follow turn from building the instrument to asking what the instrument could ever measure.

3.7 What this is and is not

What is established is twofold and solid. The regularized semiclassical count of $H = xp$ equals the smooth Riemann–von Mangoldt count of the zeros, Eq. (3.2), term for term; this is Berry and Keating's computation and it is correct. And the explicit formula, Eq. (3.3), is a theorem with the exact structure of a Gutzwiller trace formula, primes for orbits and zeros for levels. Both facts are real mathematics.

What is not established is a Hilbert–Pólya operator, and xp is not one. It supplies the average density of the zeros and not the fluctuation; the fluctuation is the primes, entering as a periodic-orbit sum whose underlying dynamical system is unknown; and the obstructions to finding that system — the absorption-spectrum sign, the chaotic-fluctuation requirement, the missing discreteness — are not technical loose ends but the fingerprints of a chaotic system that no one has constructed. The trace formula is a true and beautiful identity, and it is not a derivation of the zeros from the primes, because chaos makes the identity self-referential. Book One showed that the zeros keep the statistics of a chaotic spectrum. This chapter shows that the explicit formula is the trace formula such a spectrum would obey — and that having the trace formula without the system is exactly as far as the operator dream has come.

Reproducing the figure

```
import numpy as np
from mpmath import zetazero
g = np.sort(np.array([float(zetazero(k).imag) for k in range(1, 41)]))
T = np.linspace(0.5, 100, 2000)

# smooth (Berry-Keating / Riemann-von Mangoldt) count = xp's semiclassical density
def Nsmooth(t):
    return t/(2*np.pi)*np.log(t/(2*np.pi)) - t/(2*np.pi) + 7/8

# Panel A: smooth curve vs actual staircase  (number of gamma_n <= T)
count = np.searchsorted(g, T, side="right")
# Panel B: the fluctuation that xp does NOT supply -- the prime/periodic-orbit term
fluctuation = count - Nsmooth(T)

print(round(float(Nsmooth(100)), 3), int((g <= 100).sum()))   # 29.002  29
```

Chapter 4

An Algebra in the Primes

The natural operator from the primes had the wrong spectrum, and the natural classical Hamiltonian had only the average. This chapter follows the impulse to go bigger — to build not a single operator but a whole algebra of them from the prime data, and to hope that somewhere in its structure the zeros are hiding. The algebra is real and it is beautiful. The prime-mode bilinears close into the rank-two symplectic algebra $\mathfrak{sp}(4, \mathbb{R})$*; its natural representation splits into Dirac's two famous singletons; and because* $\mathfrak{sp}(4, \mathbb{R})$ *is the same Lie algebra as the symmetry algebra of a small curved spacetime, one can dress the entire construction as a little universe living inside the primes, complete with a field, a stress tensor, and a gesture at the cosmological constant. None of that dress is mathematics, and the chapter is careful to say which is which. The test that matters is unmoved: the algebra's own Hamiltonian still has the logarithms of the integers for a spectrum, not the zeros. The chapter ends by scoring every candidate of Part II against the two-part test, and by placing the two programs — Connes' and de Branges' — that try to clear it for real.*

4.1 Scope

This chapter is part construction, part exposition, part warning. The construction is the operator algebra generated by the prime-mode bilinears of the previous chapters, carried up to its full closed form. The exposition is the representation theory that algebra turns out to carry, which is classical and which we credit to its discoverers. The warning concerns what that representation theory does and does not mean: it is genuine structure, and it is also the most inviting place in the whole subject to mistake structure for progress, because the algebra coincides with one that physics has every reason to find interesting. The chapter keeps the mathematics and the physical costume strictly apart. It then steps back to the two serious functional-analytic programmes aimed at the operator dream, Connes' and de Branges', and reports each at its true status. As before, nothing here proves anything about the zeros, and the closing section says so plainly.

4.2 The algebra the primes generate

A word first on what is being built, in plain terms. An *algebra* of operators is a collection closed under the operations one can perform on them—here, taking commutators, the operator measure of how far two operations fail to commute. To say the collection *closes* is to say one never escapes it: combine any two members and the result is a third already in the set. Building one from the primes means asking for the smallest such set the two prime modes generate, and the answer turns out to be a named object carrying a great deal of structure—none of which mentions the zeros, which is the whole point.

Recall the two distinguished modes of the previous chapters: the prime-side spectral function and its complement, the operators we have been calling $\widehat{S}_\psi$ and $\widehat{S}_{\psi^\dagger}$, built as weighted sums over the primes in two complementary families of residues. Each is a single creation-and-annihilation object on the bosonic Fock space of prime modes. From two such modes one can form Hermitian bilinears — products of one creation or annihilation operator with another — and the question is what algebra those bilinears generate under commutation.

The answer is clean. The independent Hermitian bilinears of a two-mode system number ten, and under commutation they close: the commutator of any two is again a combination of the ten. The algebra they form is $\mathfrak{sp}(4,\mathbb{R})$, the real symplectic Lie algebra of rank two and dimension ten. This is not an analogy or an approximate match; the bilinears satisfy the defining commutation relations exactly, and one can write down a Cartan–Weyl basis — two commuting generators and the four root directions with their conjugates — entirely in terms of the prime-mode operators. The four observables of the earlier chapters sit inside this algebra as four of its ten elements; the remaining six are the squeezing and rotation generators that mix the two modes. The primes, in short, generate a copy of $\mathfrak{sp}(4,\mathbb{R})$, by hand, with no reference to the zeros at all.

4.3 Two famous representations

An algebra is one thing; how it acts is another. That "how it acts" has a name—a *representation*, a way of realizing the abstract generators as concrete operators on a concrete space of states—and the space the prime modes supply carries a famous one. The two-mode Fock space on which these bilinears act carries the *oscillator representation* of $\mathfrak{sp}(4,\mathbb{R})$ — the same representation, built from quantized oscillators, that Segal, Shale, and Weil constructed in general. It is reducible, and it splits in the simplest possible way: by the parity of the number of quanta. The even-quantum states form one irreducible piece, the odd-quantum states the other. In the prime construction these are exactly the symmetric and antisymmetric combinations of the two modes, $\widehat{S}_\psi + \widehat{S}_{\psi^\dagger}$ and $\widehat{S}_\psi - \widehat{S}_{\psi^\dagger}$, that the earlier chapters met as the even and odd sectors.

These two irreducible pieces have names, and they are famous ones. They are the *singletons* that Dirac singled out in 1963 and that Flato and Fronsdal later called the

Rac and the *Di*: two representations so small that they have no analogue in ordinary quantum mechanics, each describing a single irreducible tower of states. The Rac is the scalar singleton, with lowest weight $\Delta = \frac{1}{2}$; the Di is the spinor singleton, with lowest weight $\Delta = 1$. The even and odd sectors of the prime two-mode gas are a copy of the Rac and the Di. This is correct, and it is the kind of fact that makes the subject feel as though it is about to open onto something enormous.

4.4 The isomorphism, and the costume

The reason it feels that way is a genuine coincidence of Lie theory. $\mathfrak{sp}(4,\mathbb{R})$ is isomorphic, as a Lie algebra, to $\mathfrak{so}(2,3)$ — one of the small "accidental" isomorphisms among the low-rank classical algebras. And $\mathfrak{so}(2,3)$ is not just any algebra: it is the symmetry algebra of four-dimensional anti-de Sitter space, the curved spacetime of the same name, and equivalently the conformal algebra of its three-dimensional boundary. The Rac and the Di, in that reading, are fields living on the boundary of a small universe. So the prime two-mode gas can be dressed, step by faithful step, as a physical system: a spacetime, a boundary, a Weyl field on that boundary, a Lagrangian for the field, a stress-energy tensor, an energy density — and, with one more assumption about counting degrees of freedom on a cosmological horizon, even a number of the rough size of the observed cosmological constant.

Every arrow in that chain is a real piece of representation theory or differential geometry. And the chain is a costume. The mathematics is this: a ten-dimensional Lie algebra, generated by prime-mode bilinears, acting on a Fock space through a representation that splits into two irreducibles. The spacetime, the field, the horizon, and the cosmological constant are physical names laid over that mathematics; they are suggested by the isomorphism $\mathfrak{sp}(4,\mathbb{R}) \cong \mathfrak{so}(2,3)$ and they are not consequences of the primes. The honest version of the claim keeps only what is derived — the algebra, its representations, the proportionalities among the prime-mode quantities — and refuses the identification with physical space, physical mass, or physical vacuum energy. It is worth saying that this restraint is not a reluctance imposed from outside; the construction's own internal accounting reaches the same verdict, marking the mass as dimensionless and not a particle mass, and the geometry as not identified with the spacetime we live in. A beautiful isomorphism is an invitation to a metaphor, not a derivation of a world.

Producing a striking number. The algebra is generous with numbers, and some of them are striking. Build two real observables from the prime two-mode quantities — a quadratic and a quartic combination of the mode amplitudes, each averaged with the windowed difference of Part III, evaluated at the zeros and again at the midpoints between them — and measure how they move together. At the canonical split modulo 210, with primes counted to $X = 10^5$, the two mode energies come out as $E_+ = 0.16513$ and $E_- = 0.15251$, and the covariance of the two observables is

$$\mathrm{Cov}(Q_2, Q_4) \;=\; -0.82040.$$

A correlation of -0.82 is not a faint signal. It is a near-total anti-alignment, the kind of number that looks like a conservation law: as one observable rises the other falls, almost exactly, as though something were being held fixed between the two modes.

Dressing it as relativity. Here the costume goes on, and for a reader trained in physics it goes on almost by itself. Stack the two mode amplitudes into a single two-component object, $\psi = (a, b)^\top$, and read it as a Weyl spinor — the basic object of relativistic field theory. Contract it with the Pauli matrices $\sigma^\mu = (\mathbb{1}, \sigma^x, \sigma^y, \sigma^z)$ to form a current $J^\mu = \psi^\dagger \sigma^\mu \psi$. Because the two modes are carried on disjoint residues the spatial components J^x and J^y vanish identically, and what survives is

$$J^0 = E_+ + E_-, \qquad J^z = E_+ - E_-.$$

Now take the Minkowski length of this current, contracting with the metric $\eta = \mathrm{diag}(+,-,-,-)$:

$$m^2 := J^\mu J_\mu = (J^0)^2 - (J^z)^2 = 4\,E_+E_-.$$

Call m a rest mass, J^0 an energy and J^z a momentum, and the relation $E^2 = m^2 + p^2$ — the relativistic dispersion relation, the most recognizable equation in the subject — holds to machine precision across every modulus and cutoff one cares to test. Push further and the mass acquires a closed form: in the limit of large primorial moduli the density factor ties, through Mertens' theorem, to the twin-prime constant C_2, giving

$$m_\infty \sim \frac{4\,C_2\,e^{-\gamma}\,S(\infty)}{\log p_k} = \frac{0.770\ldots}{\log p_k}, \qquad S(\infty) = \frac{4\,\mathrm{Cin}(\pi) - \mathrm{Cin}(2\pi)}{8} = 0.519\ldots$$

Here Cin denotes the entire cosine integral, $\mathrm{Cin}(x) = \int_0^x \frac{1-\cos t}{t}\,dt$. A spinor, a conserved current, a Lorentz-invariant mass, a dispersion relation, and a clean closed-form mass with one of the famous constants of prime theory sitting inside it. It is, I will admit plainly, the most convincing thing in either book. When I first assembled it I spent the better part of a day believing the primes had handed me special relativity.

Taking the costume off. They had not. Every equation above is true, and not one of them is physics. The dispersion relation holds to machine precision because it is an identity, not a measurement: $(J^0)^2 - (J^z)^2 = (E_+ + E_-)^2 - (E_+ - E_-)^2 = 4E_+E_-$ is the difference-of-squares rule, true of any two numbers whatever. The Minkowski minus sign that makes it look Lorentzian was put in by hand, the moment we chose to contract with $\eta = \mathrm{diag}(+,-,-,-)$ rather than with the ordinary dot product; the "spinor" is two real energy sums; the spatial components vanished by construction, not by any dynamics. The mass is, in a word, kinematically thin: it is built from the mode energies alone and carries no information whatever about the zeros — the very objects the chapter set out to reach. The one genuinely arithmetic ingredient, the closed form, is honest but bare: the specific number 0.770 is fixed by the choice of window and the choice of residue split, both conventions of the construction and neither a physical input. And the striking -0.82 does not even hold still — recomputed at $X = 10^4$ and

$X = 10^6$ it reads -0.76 and -0.70. It is the value the finite computation happens to sit at in this regime, set by which residues are small primes, not a constant of anything.

So the relativity was a translation, not a discovery. The isomorphism $\mathfrak{sp}(4, \mathbb{R}) \cong \mathfrak{so}(2,3)$ is real, the Weyl-spinor contraction is legitimate notation, and the dressing is internally flawless — which is precisely what makes it dangerous. A Lie algebra appearing in the primes is not a spacetime arriving; a minus sign one has inserted oneself is not a discovery that the world is Lorentzian; and a number that drifts with the cutoff is not a law. This is the warning the chapter is really for, and it is sharpest for the reader most fluent in the costume: the more relativity you know, the more readily this assembles itself, and the more carefully you must ask which of its symbols you put in and which the primes actually gave you.

4.5 Does the algebra reach the zeros?

Put the costume aside and ask the only question that matters. This algebra is defined intrinsically from the primes, so it passes the independence half of the two-part test without effort. Does it pass the other half — is the spectrum of its natural Hamiltonian the zeros?

It is not. The Hamiltonian that generates time evolution on this Fock space is the same one from the operator chapter, the primon-gas Hamiltonian, and its spectrum is the logarithms of the integers, $\{\log n\}$, exactly as before. Enlarging the single operator to a ten-dimensional algebra, and recognizing two celebrated representations inside it, changes nothing about that spectrum. The algebra is richer; the spectrum is identical; and the spectrum is the wrong one. This is the sharpest form of the lesson the operator chapter began. Structure is not the obstacle and never was. One can have as much structure as one likes — a closed Lie algebra, a famous pair of representations, an entire borrowed spacetime — and still not have an operator whose eigenvalues are the zeros. Reaching the zeros is a different kind of achievement from building structure, and no amount of the second is any of the first.

Figure 4.1 scores the candidates of Part II against the two conditions. The made-to-order kernel has the spectrum and no independence; the primon gas, with or without its $\mathfrak{sp}(4, \mathbb{R})$ dress, has independence and the wrong spectrum; Berry and Keating's xp has independence and the average but not the fluctuation. The one candidate in the corner that passes both — Frobenius, whose eigenvalues are the zeros and which is defined without them — lives only over a finite field, where the previous chapters located the single world in which the Riemann Hypothesis is a theorem. Over the rational numbers the corner is empty.

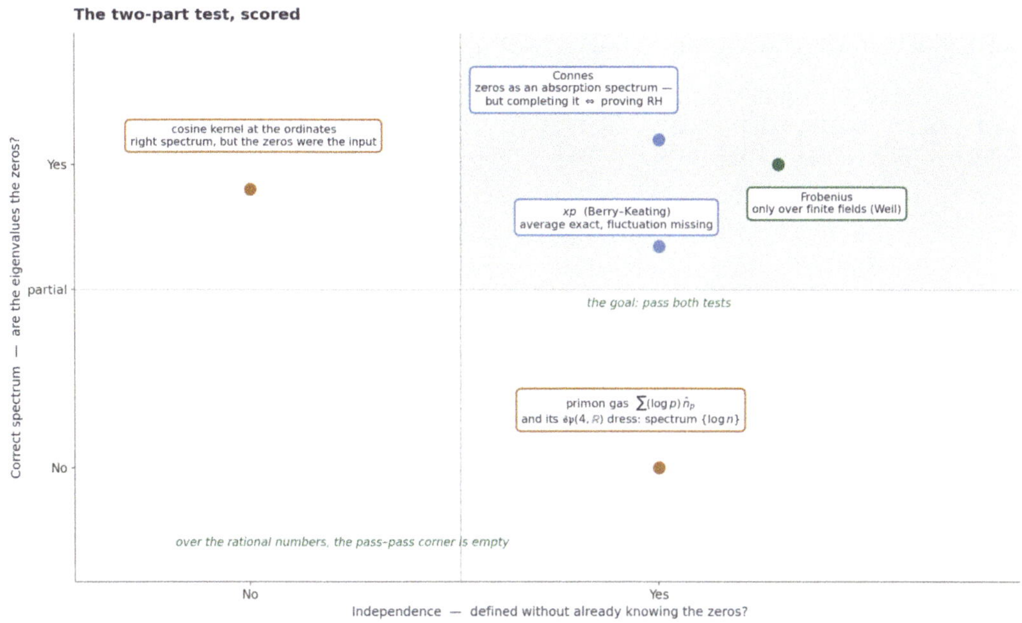

Figure 4.1: Part II against the two-part test. Horizontal: is the operator defined without already knowing the zeros? Vertical: are its eigenvalues the zeros? The made-to-order cosine kernel has the spectrum but no independence (upper left). The primon gas, and its $\mathfrak{sp}(4, \mathbb{R})$ enlargement, have independence and the spectrum $\{\log n\}$ (lower right). Berry–Keating's xp is independent and gets the average density exactly but not the fluctuating part (right, partial). Connes' construction is independent and exhibits the zeros as an absorption spectrum, but completing it is equivalent to proving the Riemann Hypothesis. The only *unconditional* occupant of the pass–pass corner is Frobenius — and only over finite fields. Over $\mathbb{Q}$ that corner is empty.

4.6 The two programmes that aim at the corner

The hand-built algebra is a toy, useful for seeing exactly how structure and spectrum come apart. Two serious programmes aim at the empty corner with the full weight of modern analysis, and both deserve their true status rather than a slogan.

Connes' programme places the problem in noncommutative geometry. On a large symmetry space attached to the rational numbers — the adele class space — there is a natural action whose spectral content can be computed, and in it the zeros of ζ appear not as emission lines but as an *absorption spectrum*: as the frequencies missing from an otherwise continuous background. This is the precise form of the minus sign noticed in the trace-formula chapter, where the primes entered as subtractions. Connes' theorem is that the Riemann Hypothesis is equivalent to the positivity of a particular functional — the Weil form of the explicit formula, read as a trace. That is a real and deep reformulation. It is also, exactly, a reformulation: the construction is independent of the zeros, and the zeros do appear in its spectrum, but establishing that the absorption spectrum is complete and sits on the critical line is equivalent to the Riemann Hypothesis itself. Connes' work moves the problem into a setting where it can be stated as a single positivity, and the positivity is as hard as the thing it would prove. On the scorecard it sits at the very corner, reachable only by an argument that would already be the proof.

De Branges' programme is functional-analytic. His spaces of entire functions — a theory powerful enough that he used it to settle the Bieberbach conjecture — come with structure functions whose positivity, in the right generality, would force the zeros of ζ and of the Dirichlet L-functions onto the critical line. The difficulty has been specific and documented: the positivity conditions de Branges required were shown by Conrey and Li not to hold in the form the argument needs, so the proposed proofs are not accepted. The machinery itself is sound and valuable; the application to the Riemann Hypothesis is unproven, and particular versions of it have been refuted. It is neither a success nor a dead end — it is an open programme with a known obstruction, which is the honest thing to call it.

4.7 What this is and is not

What is established is a chain of correct mathematics. The prime-mode bilinears close into $\mathfrak{sp}(4,\mathbb{R})$; the two-mode oscillator representation splits into the Rac and the Di singletons; $\mathfrak{sp}(4,\mathbb{R})$ is isomorphic to $\mathfrak{so}(2,3)$. Each of these is a theorem, and together they are a genuinely elegant structure sitting inside the primes. Connes' equivalence of the Riemann Hypothesis with a trace positivity, and the de Branges theory of entire-function spaces, are likewise real mathematics of a high order.

What is not established is anything about the location of the zeros. The $\mathfrak{sp}(4,\mathbb{R})$ algebra, for all its structure and its borrowed spacetime, has a Hamiltonian whose spectrum is $\{\log n\}$; it passes independence and fails the spectrum, and the famous representations do not change that. The spacetime, the field, and the cosmological constant that the isomorphism invites are a costume over the representation theory and carry no inference about the primes. Connes' programme converts the problem into a positivity that is exactly as hard as the Riemann Hypothesis; de Branges' converts it into a positivity that is known, in the proposed form, to fail. The operator dream ends Part II where the scorecard leaves it: the corner that would settle everything is occupied only over finite fields, and over the rationals the most beautiful algebra we can build from the primes still does not reach the zeros. Part III stops trying to build the instrument and asks instead what such an instrument, if we had it, could ever measure.

On the figure

Figure 4.1 is a schematic, not a computation: it places each operator of Part II by how it fares on the two conditions argued through Chapters 2–4. The one numerical anchor behind it is the spectrum of the primon-gas Hamiltonian, $\{\log n\}$, established in the operator chapter from unique factorization, and shared without change by the $\mathfrak{sp}(4, \mathbb{R})$ enlargement of this chapter.

Reproducing the number

The mode energies and the covariance are the book's own computation; the program below produces them at the canonical split modulo 210 with primes counted to $X = 10^5$ (the thousand zeta zeros take a few minutes to generate via `mpmath`). Changing X reproduces the drift the previous section describes — the -0.82 does not hold still.

```
import numpy as np
from math import gcd, log, sqrt
from mpmath import zetazero, mp

M, X = 210, 10**5

# substrate: the twin-corridor split of the wheel
Zstar = [r for r in range(1, M) if gcd(r, M) == 1]            # 48 reduced residues
Zset  = set(Zstar)
T     = [r for r in Zstar if (r+2) % M in Zset or (r-2) % M in Zset]
S_psi = {r for r in T if r < M/2}        # psi sector      (15 residues)
S_pd  = {r for r in T if r > M/2}        # psi-dagger sector (15 residues)

# primes coprime to M, with the sin^2 taper
def primes_to(N):
    s = np.ones(N+1, bool); s[:2] = False
    for i in range(2, int(N**0.5)+1):
        if s[i]: s[i*i::i] = False
    return np.where(s)[0]
primes = np.array([p for p in primes_to(X) if gcd(int(p), M) == 1])
logp = np.log(primes.astype(float))
w    = np.sin(np.pi * (logp/log(X)) / 2)**2
c    = w / np.sqrt(primes.astype(float))
res  = primes % M
inpsi = np.array([int(r) in S_psi for r in res], float)
inpd  = np.array([int(r) in S_pd  for r in res], float)
Ep = float(np.sum((c*inpsi)**2))     # E_+
Em = float(np.sum((c*inpd )**2))     # E_-

# first 1000 zeta zeros; zero-centres and midpoint-centres
mp.dps = 15
g = np.array([float(zetazero(n).imag) for n in range(1, 1001)])
K, step = 12, 23
W = (1000 - 120)//step + 1
zc = np.array([g[j*step+k]                    for j in range(W) for k in range(K)])
mc = np.array([0.5*(g[j*step+k]+g[j*step+k+1]) for j in range(W) for k in range(K)])
n  = len(zc)

# windowed-difference covariance of Q1..Q4
norm = 4.0/sqrt(Ep*Em)
cpsi = (c*inpsi).astype(complex); cpd = (c*inpd).astype(complex)
def cov(centres):
    ph = np.exp(1j*np.outer(centres, logp))
```

```
    sp, sd = ph@cpsi, ph@cpd; qp, qd = np.conj(sp), np.conj(sd)
    a, b = norm/2, norm/2j
    AB = np.array([[a*sp,a*sd],[b*sp,-b*sd],[a*qp,a*qd],[-b*qp,-b*qd]])
    CD = np.array([[a*qp,a*qd],[-b*qp,b*qd],[a*sp,a*sd],[b*sp,b*sd]])
    Am = AB[:,0:1,:]*CD[None,:,0,:]*Em + AB[:,1:2,:]*CD[None,:,1,:]*Ep
    return (0.5*np.real(Am+np.conj(Am.transpose(1,0,2)))).sum(2)
C = (cov(zc) - cov(mc))/n

print("E_+ =", round(Ep,5), "  E_- =", round(Em,5))
print("Cov(Q2,Q4) =", round(C[1,3],5))
```

Part III

Hearing and reading

Chapter 5

Hearing the Zeros

Part II could not build the instrument. This part asks what the instrument could measure if we had it, and the answer divides cleanly in two. From the primes alone one can detect *the zeros: a coherence built out of the prime data lights up, at the very floor of statistical significance, whenever it is evaluated within one mean spacing of a zero ordinate, and goes dark elsewhere. The zeros announce themselves; you can hear that they are there. What one cannot do is* resolve *them — separate two nearby zeros, or read off an ordinate to better than the data allows — and this is not for want of trying. The sharpest modern line-finding methods, ESPRIT and atomic-norm super-resolution, applied honestly to the prime-side signal, do no better than the crude periodogram, and in one case slightly worse. The reason is a single limit that ties this chapter to the two before it: the resolution is set by the aperture, the aperture is set by the largest prime in hand, and the fine structure of the zeros sits below it. You can hear the zeros. You cannot read them.*

5.1 Scope

This chapter is part exposition and part a set of honest measurements. It assumes Book One's diffraction picture — primes and zeros as Fourier duals — and the resolution limit hinted at in this book's first chapter, where seeing a symmetry type required not seven L-functions but thousands. The new content is the clean separation of two operations that are easy to confuse: detecting that a spectral line is present, and resolving where it is. The chapter reports one positive result, that a prime-side coherence detects the zeros under a strict null control, and four negative results, that the standard super-resolution methods do not resolve them. It makes no claim about the Riemann Hypothesis: the zeros are computed independently and used only as ground truth, to score detection and resolution against. Section 5.6 states the boundary the chapter draws, which is the resolution wall in the language of signal processing.

5.2 Two questions, easily confused

When a signal contains a pure tone, two different things can be asked of it. The first is *detection*: is there a line near this frequency at all? The second is *resolution*: exactly where is it, and if there are two lines close together, can they be told apart? In ordinary listening the two run together, but they are governed by different limits. Detection needs only that the line carry enough energy to rise above the noise. Resolution needs that the observation be long enough — the aperture wide enough — for two nearby lines to separate. A short recording of a chord tells you a chord is there (detection) without letting you name the two closest notes (resolution). The whole of this chapter is the claim that, for the zeros read off the primes, these two come apart: detection succeeds and resolution fails.

5.3 You can hear them

Build, from the primes in two complementary families of residues, the two prime-side signals of the earlier chapters, and form a coherence between them — a measure of how strongly the two move together — as a function of the height t at which they are evaluated. Now ask whether that coherence knows about the zeros. It does, and sharply.

Restricted to heights within one mean spacing of a zero ordinate, the coherence is large and statistically unmistakable. Tested against a null model that randomly flips the signs of the contributing terms — a Rademacher null, run for twenty thousand replications — the observed coherence is more extreme than every single replicate, pinning its significance at the Monte Carlo floor, a p-value of 5×10^{-5}. Restricted instead to heights more than one mean spacing away from any zero, the same coherence collapses to the null: it is indistinguishable from the sign-flipped surrogates. The signal is null-controlled and grid-replicated, so the effect is not an artifact of one window or one character. In plain terms: the primes, combined this way and listened to at the heights of the zeros, light up; listened to anywhere else, they are silent.

This is the explicit formula of Book One made audible. Primes and zeros are Fourier duals, so a prime-side quantity evaluated at the dual frequencies — the zero ordinates — interferes constructively, and the coherence is the sound of that interference. It is a genuine detection of the zeros from the prime side, and it is exactly as much as the prime side will give.

5.4 You cannot read them

Detection invites a hope: if the coherence rises near a zero, perhaps a sharp enough method can turn the rise into a precise location, or split two zeros that lie close together. The hope is natural, and over a careful study it fails four times.

First, separating the prime-side signal into orthogonal channels — demixing, the obvious move to clean up the data — does not buy resolution. Across thirty-six configurations, sweeping characters, window lengths, and model orders, the standard subspace method ESPRIT applied to the *undemixed* signal resolved 52.5% of close pairs, while ESPRIT on the demixed channel resolved only 42.5%. Demixing slightly *hurt*. It is a tool for telling the contributions of different L-functions apart, not for telling two nearby zeros apart.

Second, exploiting the wheel's mirror symmetry, which halves the effective bank of corridors, reduces the cross-talk between opposite-parity characters by a mere 3.7 percentage points and yields no resolution advantage at all.

Third, the most celebrated modern super-resolution method — atomic-norm minimization, the convex approach of Tang, Bhaskar, Shah, and Recht that provably beats the classical limit under the right conditions — failed to converge to a usable answer on the prime-side problem, returning degenerate spectra even on a synthetic test signal where it is known to work. The honest reading, and the one the study records, is that we cannot tell a fundamental obstruction from a solver shortfall; either way, no resolution was gained.

Fourth, stacking many channels and reweighting them by prime-gap information — gap-weighted multichannel ESPRIT — does not resolve close pairs either, and the reason is the one that matters most. The channels carry different amplitudes, but they all share the same finite-window basis, and so they share the same systematic errors. Adding channels that all look through the same window cannot widen the window.

5.5 The aperture

That last reason is the whole chapter. Resolution is set by the aperture — the span over which the signal is observed — and for a spectrum read off the primes, the aperture is fixed by the largest prime in hand: it is roughly the logarithm of that prime. Two lines separated by less than the inverse aperture merge into a single bump; no amount of cleverness applied to the same finite data widens the window that produced it.

Figure 5.1 makes the limit visible on two zeros placed close together. Observed through the short aperture the primes actually provide, the two produce a single rounded peak: the coherence detects that there is spectral content here, and cannot say there are two lines rather than one. Observed through an aperture wide enough to separate them, the two stand apart as clean peaks — but that aperture would require primes out to astronomical size, far beyond any computation. The super-resolution methods of the previous section are precisely the attempts to fake the wide aperture from the narrow data, and on this signal they do not succeed.

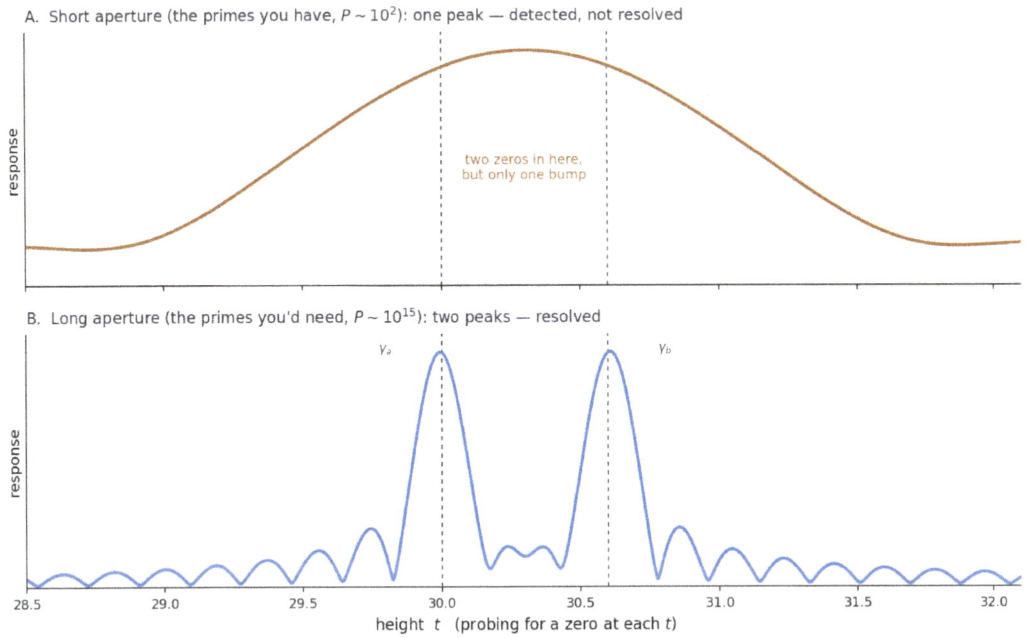

Figure 5.1: Detection without resolution. Two nearby zeros, probed by a prime-side coherence. **A:** through the short aperture the available primes provide, the two merge into one bump — their presence is detected, their separation is not resolved. **B:** through an aperture wide enough to resolve them, they appear as two peaks — but that aperture needs primes of astronomical size. The dashed lines mark the two true ordinates. Resolution is an aperture-limited quantity, and the aperture is the logarithm of the largest prime in hand.

This is the same wall met twice before, now in the language of signal processing. The first chapter needed thousands of L-functions, not seven, to see a symmetry type, because a density cannot be read from too few samples. The trace-formula chapter found a chaotic spectrum whose individual levels have no closed form and cannot be extrapolated. Here the same limit appears as an aperture: the average, coherent behaviour of the zeros is available to detection, and their individual fine positions sit below the resolution floor set by the finite prime data. The three descriptions — too few samples, no closed form, too narrow an aperture — are one fact wearing three coats.

5.6 What this is and is not

What is established is a detection and four non-resolutions, all measured. The prime-side coherence detects the zeros within one mean spacing, at the floor of a twenty-thousand-replicate null; this is a real, reproducible, null-controlled effect. The four resolution attempts — demixing, mirror symmetrization, atomic-norm minimization, gap-weighted multichannel ESPRIT — are honest negative results, reported as such, including the one where method and solver cannot be cleanly separated. None of this touches the Riemann Hypothesis. The zero ordinates were computed independently to high precision and used only as the ground truth against which detection and resolution were scored; nothing here certifies a zero on the critical line or bounds one away from it.

What this is, finally, is the line between hearing and reading. You can confirm, from the primes alone, that a zero sits at a given height — the coherence rises and the null rejects chance. You cannot, from the same primes, resolve two zeros that lie within an aperture of each other, or predict where the next one falls. The same chaos that gives detection a coherent signal to latch onto is what denies resolution a shortcut: a chaotic spectrum is detectable in the average and unpredictable in detail. The instrument Part II could not build would, even if built, face this second wall. The next chapter turns to what happens when one forgets the wall and mistakes a detected pattern for a law.

Reproducing the figure

```
import numpy as np

# two zeros placed close together near height 30
w1, w2 = 30.0, 30.6

# matched-filter (periodogram) response of a finite-aperture observation:
#   aperture L = log(largest prime).  short = the primes you have;
#   long  = the primes you would need to separate the pair.
def response(L, probe):
    tau = np.linspace(0, L, 6000); dt = tau[1] - tau[0]
    s = np.cos(w1*tau) + np.cos(w2*tau)
    return np.array([abs((s*np.exp(-1j*w*tau)).sum()*dt) for w in probe])

probe = np.linspace(28.5, 32.1, 1400)
short = response(5.0,  probe)   # largest prime ~ e^5  ~ 1.5e2  -> one bump
long  = response(34.0, probe)   # largest prime ~ e^34 ~ 6e14   -> two peaks
```

The figure isolates the aperture limit. The resolution percentages and the complement-symmetry figure quoted earlier in the chapter come from a longer sweep across characters, window lengths, and model orders; that sweep, and the symmetry test, are reproduced by the verification programs accompanying this book.

Chapter 6

Laws That Died

The previous chapter showed that the primes detect the zeros: patterns are there to be found. This chapter is about the trap that opens the moment a pattern is found, which is to mistake it for a law. A pattern is whatever the finite data happens to show; a law is what survives being pushed. The difference is measured by pushing, and most apparent spectral laws do not survive. This chapter is a catalogue of ones that did not: a clean cross-modulus relation that vanished when the prime cutoff was raised; a rule that held to a prime cutoff of a billion and then broke the instant the modulus was lifted to the next wheel; a ranking that inverted depending on which of two observables was used to read it; a contribution ruled out because it came up a thousandfold too small; a celebrated pattern from elsewhere in the subject that simply was not present when looked for; and a near-perfect correlation that turned out to be an artifact of which residues happen to be small primes. None of these is a failure to be hidden. The running, numbered tally of them is the most honest instrument the subject has, and keeping it is what separates the work from numerology.

6.1 Scope

This chapter is a set of negative results, each reported as a measurement. It assumes nothing beyond the observables of the earlier chapters and the wheel of Book One. Its content is six patterns that looked like laws and were killed by one or another standard test — raising the prime cutoff, lifting the modulus, changing the observable, checking the magnitude, looking for a known pattern, asking whether a striking number is structure or artifact — together with the discipline of recording such deaths in a numbered list. The chapter proves nothing new; that is the point. It is the inventory of what the methods of this book throw up that does not hold, and Section 6.9 draws the lesson for the three walls: they are hard in part because the false laws on the way to them are so plausible.

6.2 A relation that died when the cutoff rose

The simplest death comes from raising the prime cutoff. Working at primes up to 10^5, one finds a clean numerical relationship between observables computed at two different moduli — a cross-modulus shape similarity, agreeing to within a few percent, exactly the kind of tidy fact one would want to promote to a structural law. Raise the cutoff to 5×10^5 and the relationship is gone. It was a feature of the smaller sample, not of the integers, and the honest record marks it as retracted rather than burying it. The first test any candidate law must pass is that it not depend on how far up the primes you happened to count, and this one failed it.

6.3 A rule that survived a billion and died at the next wheel

The second death is sharper, because the pattern survived a far more demanding version of the first test and died on a different axis entirely. It is also, on the wheel of 210, genuinely beautiful, which is what makes its death worth telling.

Recall from Book One that the seven real characters of the wheel are indexed by their conductors, and that every conductor is a product of a subset of the three odd primes $\{3, 5, 7\}$: conductor 7, conductor $15 = 3 \cdot 5$, conductor $105 = 3 \cdot 5 \cdot 7$, and so on through all seven non-empty subsets. The conductors form a Boolean algebra — the lattice of subsets of $\{3, 5, 7\}$ — and one can ask whether the *sign* of the spectral observable is a Boolean function of which primes a conductor contains. It is, and a strikingly simple one. Writing x_p for "the conductor contains p," the observable is positive exactly when

$$x_5 \oplus x_7 = 1, \tag{6.1}$$

that is, when the conductor contains exactly one of 5 and 7. The prime 3 plays no role at all; it only pairs the characters within each cell of the rule. Figure 6.1 is the rule as a truth table, with each of the seven characters in its place.

This is exactly the kind of clean statement one wants to promote to a law, and it was tested as one. The rule was written down and time-stamped *before* the decisive run, then checked at a prime cutoff of 10^9 — several thousand zeros per character, hours of computation each — and it held, seven characters out of seven. One character, of conductor 21, sat on a knife-edge: its observable was the smallest of all seven and had changed sign at lower cutoffs, so the sign-only prediction was caught correctly, but by a hair. Another, of conductor 35, was the lone character never to change sign across every cutoff from 10^4 to 10^9. A pre-registered parity law, clean to a billion, is the sort of evidence that ordinarily settles a question.

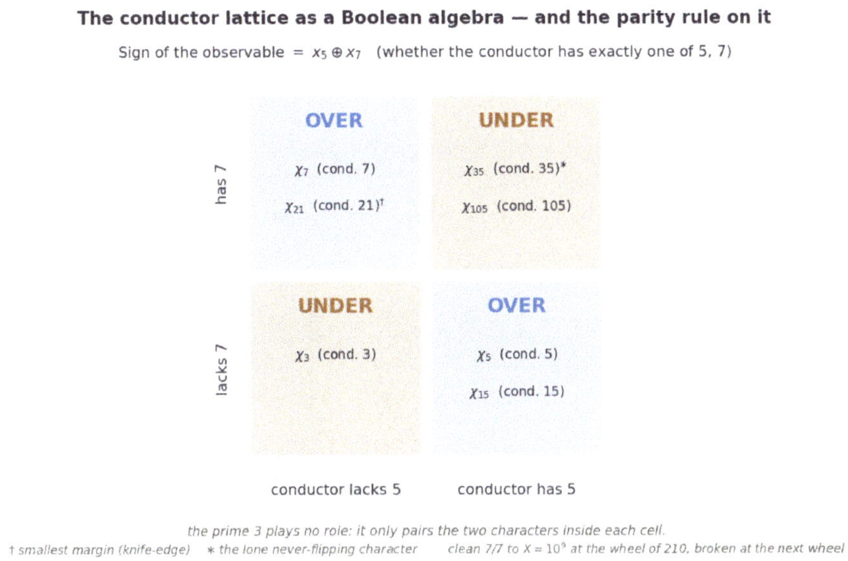

Figure 6.1: The parity rule on the conductor lattice of the wheel of 210. The seven characters are the seven non-empty subsets of $\{3, 5, 7\}$; the sign of the observable is positive (OVER) exactly when the conductor contains exactly one of 5 and 7, Eq. (6.1). The prime 3 is irrelevant: it only pairs the two characters inside each cell. Clean across all seven characters to a prime cutoff of 10^9 — and broken at the next wheel.

Then lift the modulus. Carry the same rule, transcribed literally, from the wheel of 210 to the next wheel, $2310 = 2 \cdot 3 \cdot 5 \cdot 7 \cdot 11$, and it breaks. It does not weaken or blur; it fails outright, on several of the original characters and on the eight new ones the larger wheel introduces. A brute-force search over the fifteen characters of the new wheel for the best simple Boolean replacement turns up a different, weaker rule — positive unless the conductor contains 7 — which fits thirteen of fifteen, with two stubborn exceptions, and which is significant only after correcting for the handful of simple rules one tried, not against the full space of possibilities. The clean parity law of the wheel of 210 has become, at the wheel of 2310, a rough rule of thumb. (Stranger still, the wheel of 1155, which is 2310 with the factor of two removed, behaves almost identically, the two agreeing to within a few parts in ten thousand — so the failure tracks the odd part of the modulus, not the size.)

The lesson is that the prime cutoff and the modulus are two independent axes of stress, and surviving one is no guarantee of surviving the other. A pattern can be true for every prime up to a billion on one wheel and false on the next. Holding to 10^9 proved only that the rule was not an artifact of small primes; it said nothing about whether the rule was an artifact of the wheel 210 itself, and it was.

6.4 A verdict that inverts

The third death needs no new data at all — only a second way of reading the same data. Throughout these chapters a "canonical" arrangement of the residues has been compared against a thousand random arrangements, and rated by where it falls in the random distribution: a high percentile means the canonical arrangement is special, a percentile near the middle means it is ordinary. The verdict depends, it turns out, on which observable does the rating.

Figure 6.2 shows two observables — a signed bilinear quantity and its absolute-value cousin — rating the same canonical split. On the wheel of 210 they agree: both place it high, at the 99.9th and 92.8th percentiles, both calling it special. On the wheel of 2310 they invert. The signed observable rates the canonical split at the 100.0th percentile, above all thousand random arrangements; the absolute-value observable rates the identical split, on the identical data, at the 43.7th percentile — below the random average. One observable says this arrangement is more extreme than anything chance produces; the other says it is slightly less special than a coin flip. A property that flips from "extraordinary" to "below average" depending on how you choose to measure it is not a property of the arrangement. It is a property of the measurement, and a candidate law that survives only under one of two equally natural observables has already failed.

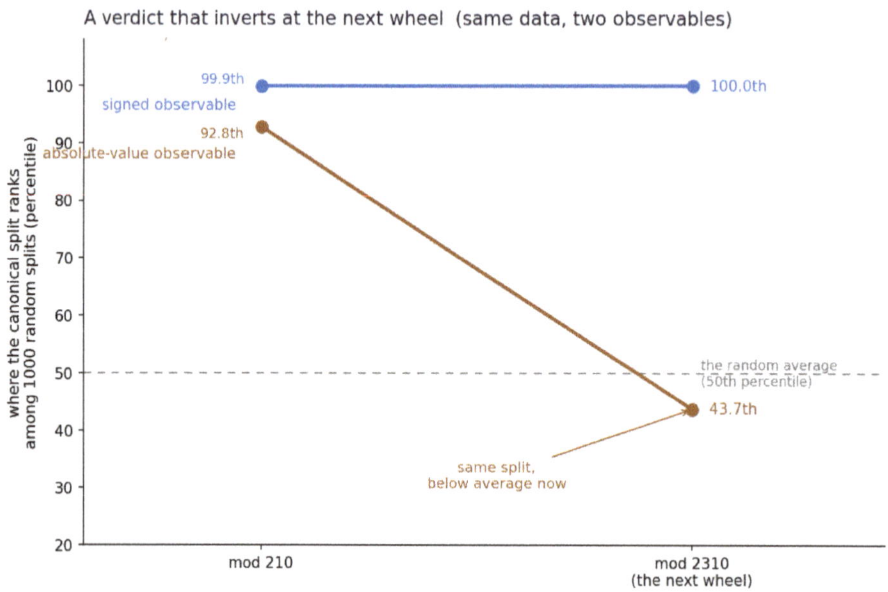

Figure 6.2: The same canonical split, rated by two observables, on two wheels. On the wheel of 210 both rate it special (top left). On the wheel of 2310 the signed observable still rates it at the very top (100.0th percentile) while the absolute-value observable drops it below the random average (43.7th). Same data, opposite verdict: the ranking is a fact about the observable, not about the split.

6.5 An idea killed by its size

The fourth death is the cleanest, because it is not subtle. One of the most natural ways to try to explain a residual gap in the data is to attribute it to the archimedean factor — the gamma-function piece and root number that complete an L-function at the infinite place. It is a real contribution and an obvious suspect. So one computes it, for all seven real characters of the wheel, and compares its size to the gap it is meant to explain. It is more than three orders of magnitude too small. There is no fitting to be done and no parameter to adjust; the candidate explanation is simply the wrong size by a factor of thousands, and it joins the running list of tested-and-falsified hypotheses — a magnitude failure, a different kind of death from the change-of-wheel collapses tallied later in the chapter. Not every death is a subtle reversal at scale. Some are a number that is off by a thousandfold, and the discipline is to check the magnitude before believing the story.

6.6 A pattern that was not there

The fifth death is of a pattern that belongs to someone else. In 2022 a striking correlation was discovered by He, Lee, Oliver, and Pozdnyakov, using machine learning, in the data of elliptic-curve L-functions: averaged over a family and plotted against the prime p, for curves in a fixed conductor window, the coefficients trace out a faint, coherent, oscillating shape, since named a murmuration. It is a real and now much-studied phenomenon, and its discovery is exactly the kind of family-wide structure the methods of this book are built to look for. So one looks, in the families natural to the wheel. The murmuration is not there. The honest record states it plainly — no such correlation is found, and no claim is made that the wheel's families murmurate. Looking for a beautiful pattern known to exist elsewhere, and reporting its absence here, is as much a result as finding one, and a good deal rarer in practice.

6.7 A number that was an artifact

The sixth death was promised two chapters ago. The hand-built algebra of the operator part threw up a striking number: a correlation of about -0.82 between two of its observables, a near-total anti-alignment that looks for all the world like a conservation law binding the two modes. Pushed, it dissolves. The two channels of that algebra are built from two complementary families of residues, and the family on one side consists of residues that are themselves small primes — 11, 13, and their kin — while the family on the other side contains no small prime at all below one hundred. The strong anti-alignment is a consequence of that lopsided arithmetic: one channel is carried by the smallest, heaviest-weighted primes and the other is not. Change which residues fall on which side — lift the modulus, or redraw the families — and the -0.82 moves. It was never a law coupling two abstract observables; it was a shadow of which numbers happen to be small primes. A striking constant is the most seductive kind of pattern, and the test it must pass is whether it stays put when the arithmetic underneath it is changed.

6.8 Keeping a body count

Six deaths, six different tests: the prime cutoff, the modulus lift, the choice of observable, the magnitude, the search for a known pattern, the change of arithmetic. What ties them together is not that the methods are weak — they are the same methods that detected the zeros in the previous chapter — but that finite data over a small modulus will always show patterns, and the chaotic richness of the zeros guarantees that many of those patterns will look meaningful. The only defense is to try, systematically, to kill every pattern one finds: to push the cutoff, lift the wheel, swap the observable, check the size, compare against a null, and ask whether a number is structure or coincidence. The findings that survive all of it are few, and the ones that do not deserve to be written down, in order, with their cause of death.

That is why the most honest instrument in this subject is a numbered list of tested-and-falsified hypotheses, kept openly, with retractions marked in the text rather than quietly deleted. The list is real and it is specific. The exclusive-or law of this chapter was, when it broke, the *ninth* entry in one such running tally — a tally of patterns clean on the wheel of 210 that failed to survive a change of wheel, joining a factor-two hypothesis, a seam-pair relation, a character-concentration effect, the residue pairing of the next section, and several others before it. A program that records its dead ends in order is doing the one thing that separates this work from numerology, which is distinguishing what it has shown from what it has merely seen. The instinct to promote a clean coincidence to a law is exactly the instinct that the three walls punish, and the habit of killing one's own patterns is the habit that keeps the walls in view.

6.9 What this is and is not

What is established is that each of these six patterns failed a specific, standard test, and that the failures are recorded as measurements: the cross-modulus relation did not survive a fivefold increase in the prime cutoff; the exclusive-or rule held to a cutoff of 10^9 on one wheel and broke on the next; the canonical ranking inverted between two observables on the same data; the archimedean contribution was too small by three orders of magnitude; the murmuration was absent from the wheel's families; the -0.82 correlation was an artifact of small primes. These are real results, and reporting them is not a confession but the method working as intended.

What is not established is any law. Not one of these patterns is a theorem, and the chapter is deliberately a catalogue of near-laws rather than laws. The lesson it leaves for the rest of the book is about the walls. The Riemann Hypothesis, the twin-prime conjecture, and the hardness of factoring are not surrounded by a desert in which nothing happens; they are surrounded by a thicket of plausible patterns, clean coincidences, and striking numbers, most of which die when pushed. That thicket is part of why the walls are hard: it is genuinely difficult to tell, from inside a finite computation, the pattern that is the edge of a theorem from the pattern that is the edge of nothing. The remaining chapters walk up to the three walls themselves, having learned to distrust the view.

On the figure

Figure 6.2 plots reported percentile ranks, not a fresh computation: the canonical split's standing among 1000 random splits, under the signed observable (99.9th at mod 210, 100.0th at mod 2310) and the absolute-value observable (92.8th at mod 210, 43.7th at mod 2310), all at prime cutoff $X = 10^5$. These are the measured values discussed in the text; the chart simply draws the inversion between the two wheels.

Part IV

The walls

Chapter 7

A Family of Walls

Book One ended at three walls. This part walks up to them again through the spectral lens, and the first thing the lens shows is that the Riemann wall is not one wall. The spectral signal of the wheel — the coherence at the zeros, the cancellation in the observables, the prime race of Book One — is built not from ζ alone but from the whole family of Dirichlet L-functions the wheel carries, and to control it one would need the Riemann Hypothesis for every one of them at once. That is the Generalized Riemann Hypothesis, strictly stronger than the original and exactly as open. The chapter makes the dependence concrete. A handful of subprime residues do most of the cancelling, dominated by $209 = 11 \cdot 19$; and 209 matters because the characters of conductor 11 and 19 carry the two L-functions whose zeros push hardest against ζ's. Worse for any hope of a shortcut, which L-functions oppose ζ is not even fixed: lift the wheel from 210 to 2310 and the clean split dissolves. The spectral approach does not reduce the Riemann wall to something smaller. It reveals it as a coupled, wheel-dependent family of walls.

7.1 Scope

This chapter is compressed and assumes Book One's closing chapter, where the three walls were named, and its prime-race chapter, where the race was found to wander at the mercy of the zeros of the wheel's L-functions. It does not re-teach the Riemann Hypothesis; it views the wall through the observables of this book. The content is a single concrete dependence, traced in three steps: which residues carry the spectral cancellation, which L-functions among the wheel's family move against ζ, and how both reorganize when the modulus is lifted. The conclusion is that the spectral structure of the wheel is hostage to the Generalized Riemann Hypothesis, and Section 7.8 is careful that this is a door into the wall, not a passage through it. Nothing here proves, approaches, or weakens any form of the hypothesis.

7.2 The wall is not just zeta

Book One's prime race already stood at this wall without quite naming it. The race among the residue classes wandered, rather than settling, because its fluctuations are governed by the nontrivial zeros — and not only the zeros of ζ, but the zeros of all the Dirichlet L-functions attached to the wheel's characters. The same is true of every spectral quantity in this book. The coherence that detected the zeros, the cancellations that filled the observables, the rankings that inverted in the last part: all of them are sums over the primes sorted by residue, and the behaviour of such sums is controlled, through the explicit formula, by where the zeros of the corresponding L-functions sit.

So the wall that matters for the wheel is not the Riemann Hypothesis alone. It is the *Generalized* Riemann Hypothesis: the statement that *every* Dirichlet L-function has its nontrivial zeros on the critical line. The ordinary hypothesis is the single case of the trivial character; the generalized one asks the same of the whole family at once. It is strictly stronger, and it is just as unproved. Reading the wheel spectrally does not let one trade the Riemann wall for an easier obstacle; it commits one to the harder one.

7.3 Which numbers do the cancelling

The dependence can be made concrete by asking which residues actually carry the spectral cancellation. Among the residues coprime to 210, a few are composite — subprime, in the vocabulary of these books: composite, but with no factor below eleven. The smallest of them are $121 = 11^2$, $143 = 11 \cdot 13$, $169 = 13^2$, $187 = 11 \cdot 17$, and $209 = 11 \cdot 19$. When one measures how much each residue class contributes to the cancellation in the relevant observable, the work is done almost entirely by two of these: 169 and 209. Removing the primes in class 169 alone shifts the cancellation by some seventy per cent, and per prime the class 209 carries about three times the weight of 169.

A natural first guess is that this is about size — that the residue with the largest prime factor should dominate. It is not. The reason 209 outweighs 169 is not that 19 exceeds 13 but that of the wheel's characters, the ones that matter are the characters of conductor 11 and conductor 19, and $209 = 11 \cdot 19$ is built from both of them, while $169 = 13^2$ is built from the single, weaker conductor 13. The driver is not the magnitude of the prime factors. It is which characters' L-functions the residue is made of.

7.4 Which L-functions move against zeta

That sends the question one level deeper, to the L-functions themselves. Here *the wheel's characters* means the small set of low-conductor quadratic characters $\chi_{-3}, \chi_{-4}, \chi_{-7}, \chi_{-11}, \chi_{-19}$ — a spread of internal and external conductors, 3 and 7 dividing 210 and 4, 11, 19 not — and not the seven real Dirichlet characters of Chapter 6. These five expose the internal/external split rather than exhausting the wheel: the remaining external conductors, 13 and 17 behind the subprimes 169 and 187, are left off the figure to keep it readable — 13 being the single, weaker conductor the previous section identified behind 169. Measure the correlation between each of the wheel's L-functions and ζ, across the same windowed grid of zero heights used throughout, and a clean pattern appears at the wheel of 210 (Figure 7.1, left). The characters whose conductor divides 210 — the internal ones, conductors 3 and 7 — move *with* ζ, correlations +0.70 and +0.58. The characters whose conductor does not divide 210 — the external ones, conductors 4, 11, 19 — move *against* it. And the two strongest opponents, at −0.82 and −0.77, are exactly the characters of conductor 11 and 19: the two factors of 209. The residue 209 dominates the cancellation, then, for a reason that runs all the way down to the zeros: it is the product of the two characters whose L-function zeros push hardest in the direction opposite to ζ's. The cancellation is L-function content, and the content is the joint behaviour of a particular pair of zero sequences.

This −0.82 is not the −0.82 of Chapter 4. That one—the covariance $\mathrm{Cov}(Q_2, Q_4)$ of the operator algebra—was a hand-built artifact that drifts to −0.70 as the cutoff grows; this one is a genuine correlation between an L-function and ζ, and it collapses instead to −0.49 when the modulus is lifted to 2310. Two unrelated quantities, each regime-dependent in its own way, happen to read −0.82 at the canonical wheel: the agreement is a coincidence of value, not a hidden identity—of a piece with the near-laws this chapter has been taking apart.

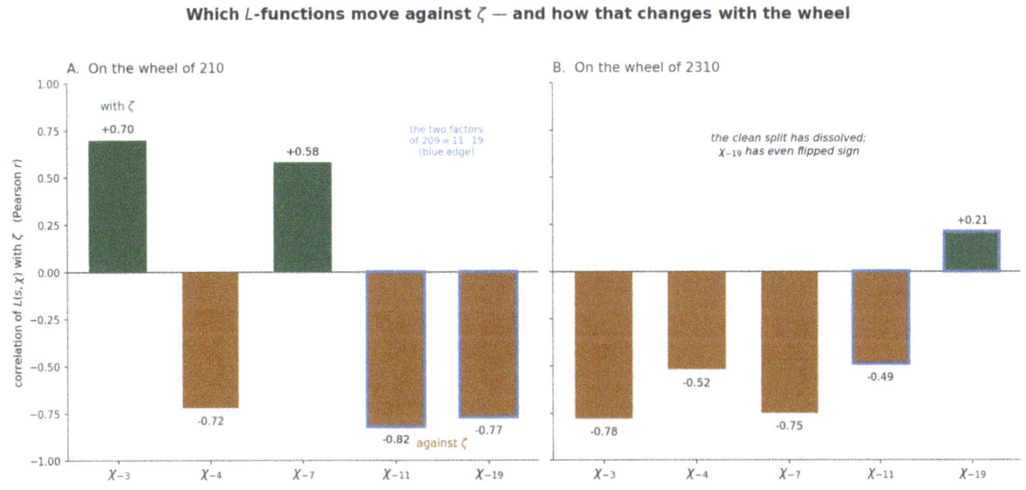

Figure 7.1: The wheel's L-functions, correlated with ζ. **Left (wheel of 210):** a clean split — internal characters (conductors $3, 7$, dividing 210) move with ζ; external characters (conductors $4, 11, 19$) move against it; the two factors of $209 = 11 \cdot 19$, outlined, are the strongest opponents. **Right (wheel of 2310):** the same correlations recomputed after lifting the modulus. The clean split has dissolved — two internal characters have gone negative and χ_{-19} has flipped to positive. Which L-functions oppose ζ is not fixed; it depends on which primes divide the wheel.

7.5 The split dissolves at the next wheel

If that pattern were stable, it would at least be a fixed target: a definite short list of L-functions to worry about. It is not stable. Lift the modulus from 210 to 2310 — adjoin the prime 11 to the wheel — and recompute the same correlations, and the clean split falls apart (Figure 7.1, right). The internal characters that moved with ζ now move against it; the conductor 11, newly internal, weakens; and χ_{-19}, the strong opponent on the wheel of 210, flips to a mild positive correlation. The organizing principle of the left panel — internal with ζ, external against — has simply dissolved.

This is the modulus-lift lesson of the previous part, arriving now at the quantities that control the Riemann wall. It is not only that a given correlation changes value; it is that the *structure* of which L-functions oppose ζ is itself a function of the wheel. There is no wheel-independent fact of the form "these characters are the opponents." Change which primes divide the modulus and the membership of the opposition changes with it, because the internal–external distinction — and with it the whole pattern — is defined relative to the wheel.

7.6 How far the explicit formula gets

So far the chapter has been diagnostic: the signal depends on a whole family of L-functions whose membership shifts with the wheel. The constructive question is sharper. How much of the observable can one actually *predict* from analytic inputs — the explicit formula, the zeros, the characters, with nothing fitted to the answer — and where does the prediction stop? The answer draws the wall in a single picture.

The observable decomposes, over the Dirichlet characters of the wheel, into four pieces supported on disjoint parity classes; the decomposition rests on an exact parity rule among the characters and the vanishing of the mixed-parity terms. One of the four pieces is special: the principal-character piece, built from the trivial character, whose L-function is ζ itself (stripped of the Euler factors the wheel removes). That piece, and only that piece, can be written directly through a Riemann-style explicit formula — an amplitude expressed as a sum over the zeros of ζ against an explicit kernel — and evaluated from the first thousand zeros. Figure 7.2 is the result of doing so.

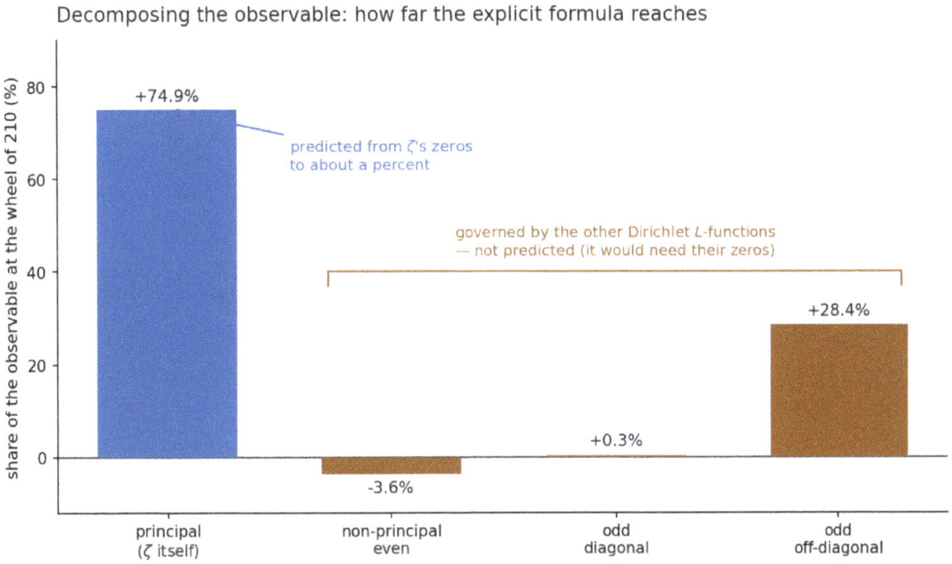

Figure 7.2: The observable at the wheel of 210, split into the four character-parity pieces. The principal piece — governed by ζ's own zeros — is about three-quarters of the whole and can be predicted from the explicit formula to about a percent. The remaining three pieces are governed by the non-principal Dirichlet L-functions of the wheel, and predicting them would require the zeros of that whole family; that prediction is not made.

At the wheel of 210 the principal piece is about three-quarters of the entire observable, and the prediction lands within roughly a percent of the measured value. The honest figure is "about a percent," not less: the pointwise prediction carries a conditional-convergence bias of several to many percent that does *not* shrink as more zeros are added, and it is the windowed difference — the same zero-minus-midpoint contrast of the detection chapter — that cancels most of that bias, so the sub-percent agreement on any one run rides on a cancellation rather than on genuine convergence. Across other wheels the principal piece is a smaller or larger share of the observable, from about a quarter to almost all of it, and the same method tracks it at the same percent level. That much is a real and non-trivial reach: most of the signal, at the canonical wheel, recovered from the zeros of ζ alone.

And then it stops, exactly where this chapter says it must. The remaining pieces — a quarter of the signal at the canonical wheel, dominated by a single off-diagonal contribution worth nearly thirty percent — are governed not by ζ but by the non-principal L-functions of the family. Predicting them would require locating and summing the zeros of those L-functions, the same family whose shifting allegiances filled the earlier sections. That prediction is left unmade, as the open problem it honestly is. The boundary between what the explicit formula delivers and what it cannot is the boundary between ζ and the rest of the family — which is to say, the boundary between the Riemann Hypothesis and its generalization. The method reaches the first wall and halts at the family.

7.7 Why this is the wall

Put the three steps together. The spectral signal of the wheel is carried by a handful of residues; those residues matter because of the L-functions they are built from; those L-functions matter because of where their zeros sit relative to ζ's; and the entire arrangement reorganizes when the wheel is lifted. To predict or control the signal one would need to know the joint placement of the zeros of all the wheel's L-functions at once — which is the Generalized Riemann Hypothesis — and one would need to know it afresh for each wheel, because the couplings are not invariant. The spectral lens does not shrink the Riemann wall to a single manageable question. It resolves the one wall into a coupled family of them, one per L-function, with the couplings themselves depending on the modulus.

That is the honest shape of the obstacle, and it is larger than the one Book One drew, not smaller. Book One's wall was the placement of the zeros of ζ. The wheel's spectral structure depends on the placement of the zeros of every L-function it carries, on how those placements correlate with one another, and on an arithmetic — which primes divide the wheel — that rearranges the correlations from one modulus to the next. The door this chapter opens leads into the wall and onto a larger room behind it. It is a clearer view of the obstruction, not a way around it.

7.8 What this is and is not

What is established is a chain of measurements. The spectral cancellation is dominated by the subprime residues 169 and 209, with 209 carrying roughly three times the per-prime weight; the wheel's L-functions correlate with ζ in the pattern shown, with the conductor-11 and conductor-19 characters the strongest opponents at -0.82 and -0.77; and that pattern dissolves when the modulus is lifted to 2310. And the principal-character piece of the observable — about three-quarters of it at the canonical wheel — is predictable from the explicit formula and ζ's zeros to roughly a percent, while the pieces carried by the other L-functions are not predicted at all. These are computed facts about the wheel's observables and the first thousand zeros, nothing more.

What is not established is anything about the Generalized Riemann Hypothesis. Nothing here bounds a zero, certifies one on the line, or moves the hypothesis an inch. The chapter shows only that the spectral structure of the wheel is hostage to GRH — that to control the signal one would need the zeros of an entire family of L-functions under simultaneous control, on each wheel separately — which makes the spectral approach a door into a harder wall rather than a shortcut past an easier one. The correlations are real and the dependence is real; the wall they lead to is the strongest form of the Riemann question, and it stands exactly where it stood. The next chapter turns to the second wall, the twin primes, where the spectral lens delivers something cleaner: not a passage through, but a single number on the far side, and an honest account of what it is.

On the figure

Figure 7.1 plots measured Pearson correlations between each L-function's windowed observable and ζ's, over the grid of the first 1000 zeros at prime cutoff $X = 10^5$. Wheel of 210: χ_{-3} +0.70, χ_{-4} −0.72, χ_{-7} +0.58, χ_{-11} −0.82, χ_{-19} −0.77. Wheel of 2310: χ_{-3} −0.78, χ_{-4} −0.52, χ_{-7} −0.75, χ_{-11} −0.49, χ_{-19} +0.21. The chart draws these values; it does not compute anything new.

Chapter 8

The Barrier and the Constant

The second wall is the twin primes, and the spectral lens does two clean things to it — neither of which is to pass it. On the near side it sharpens the obstruction to a finite, provable shape. The whole twin question localizes to fifteen corridors of the wheel; inside a corridor the relevant sieve reduces to two explicit matrices, B and M; and the centered form $B - M$ is negative semidefinite, *by an elementary identity. That is the parity barrier — the known reason sieve methods cannot, by themselves, deliver twin primes — written as the sign of a quadratic form one can compute and display. On the far side the lens produces a number. A natural invariant built from the two prime-side channels, followed up the primorials, has a closed asymptotic form in which the Hardy–Littlewood twin-prime constant $C_2 = 0.6601618\ldots$ appears, by way of Mertens' theorem. The most concrete quantity the construction yields is the very constant that would count the twin primes if there are infinitely many of them. You can compute the constant. You cannot, by computing it, earn the conjecture it belongs to.*

8.1 Scope

This chapter is compressed and assumes Book One's twin-primes chapter, where the twin corridors of the wheel were introduced and the conjecture left standing at its wall. It views that wall through the linear-algebraic and spectral instruments of this book. Two results, both honest about their limits. The first is a finite identity that exhibits the parity barrier as the negative-semidefiniteness of an explicit matrix; it is elementary and exact, and it proves nothing about the existence of twins. The second is a closed-form computation in which the twin constant emerges from a two-channel invariant; it is a genuine calculation, and it does not prove the conjecture either. Section 8.6 keeps the two firmly on the near side of the wall. As with the previous chapter, the lens clarifies the obstruction; it does not remove it.

8.2 The problem, localized

Book One placed the twin primes on the wheel and observed that a twin pair $(n, n+2)$ can occupy only certain residue corridors. Sharpened, the observation is a reduction.

Among the integers greater than 7 and coprime to 210, the residues split into forty-eight survivors, and a twin pair can sit in only fifteen of them — the twin-admissible corridors, those residues r for which both r and $r + 2$ are coprime to 210. The twin-prime conjecture is then exactly the statement that at least one of these fifteen corridors contains infinitely many twin primes. Nothing is lost in this reduction and nothing is gained; it simply puts the question where the wheel can see it, as a statement about fifteen arithmetic progressions.

The fifteen corridors are worth seeing all at once (Figure 8.1). Under the wheel's mirror symmetry $r \leftrightarrow 210 - r$ they fall into seven mirror pairs and a single self-mirror slot that wraps across the top. Thirteen of the fifteen hold genuine twin primes; the two that fail are the instructive ones. The slot $(167, 169)$ is broken because $169 = 13^2$, and the wrap slot $(209, 211)$ because $209 = 11 \cdot 19$ — in both cases by a *subprime*, a composite coprime to 210 whose least factor exceeds 7, the same rough numbers that did the cancelling in the previous chapter. The picture also shows the asymmetry the rest of the chapter is about: $(167, 169)$ and its mirror $(41, 43)$ are identical as corridors, related by the exact symmetry of the wheel, yet one holds a twin and the other does not. Admissibility is symmetric; realization is not, and nothing in the wheel decides which admissible slot is filled.

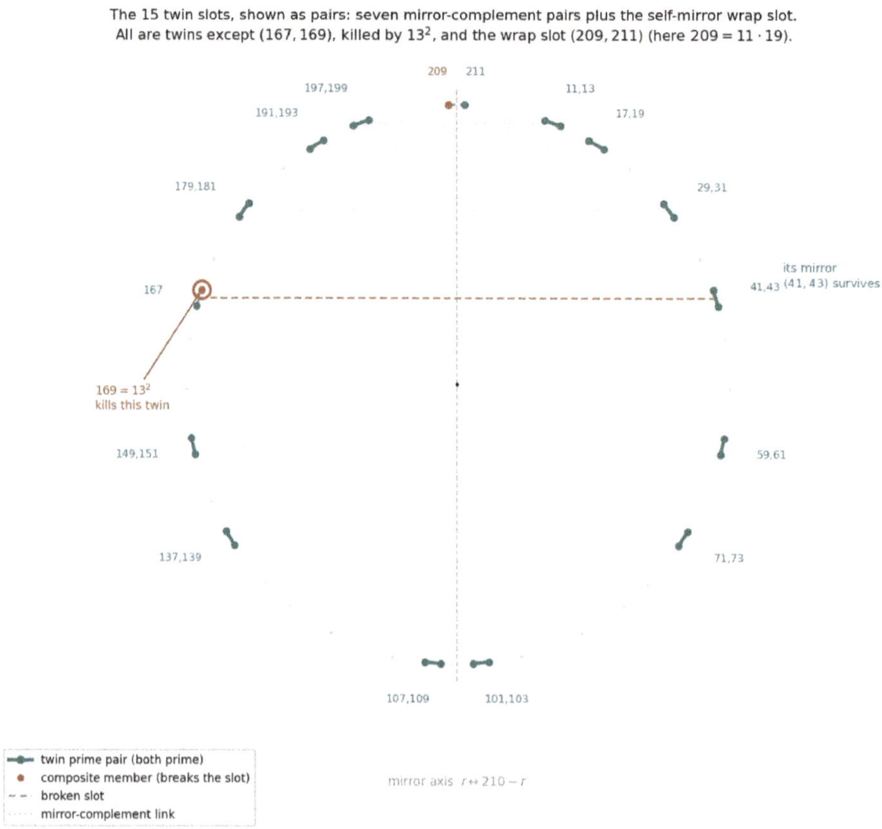

Figure 8.1: The fifteen twin-admissible corridors modulo 210. The mirror pair $(41, 43)$ and $(167, 169)$ is the chapter in miniature: identical as corridors under $r \leftrightarrow 210 - r$, related by the exact symmetry of the wheel, yet one realizes a twin and the other is broken — here by the subprime $169 = 13^2$, just as the wrap slot $(209, 211)$ is broken by $209 = 11 \cdot 19$. Admissibility is symmetric; realization is not.

8.3 Two kernels

Inside a single twin-admissible corridor, the sieve machinery for the shift pair $\{0, 2\}$ — the Selberg and Goldston–Pintz–Yıldırım variance and main terms — is governed by two matrices, indexed by squarefree moduli d coprime to 210. The first counts how often two sieve conditions overlap:

$$B(d_1, d_2) = \frac{\mathbf{1}_{\gcd(d_1,d_2)=1}}{d_1 d_2}, \tag{8.1}$$

which by the Chinese Remainder Theorem is the density of integers divisible by both d_1 and d_2 in the corridor — and which is *sparse*, carrying a coprimality indicator, precisely because 2 divides 210 while the d_i do not. The second is the smoother form

$$M(d_1, d_2) = \frac{1}{\operatorname{lcm}(d_1, d_2)} = \frac{\gcd(d_1, d_2)}{d_1 d_2}. \tag{8.2}$$

These two are the entire arithmetic content of the corridor sieve for twins; everything the method can do is built from how a chosen vector of sieve weights $\lambda = (\lambda_d)$ pairs against B and against M.

8.4 The centered identity and the barrier

The decisive comparison is between the two. Form the centered quadratic form $\lambda^\top(B-M)\lambda$ — the difference between what the overlap count gives and what the smoother form gives — and it has an exact, elementary closed expression. Using the two Möbius identities $\mathbf{1}_{\gcd=1} = \sum_{k|\gcd}\mu(k)$ and $\gcd = \sum_{k|\gcd}\varphi(k)$, one finds

$$\lambda^\top(B-M)\,\lambda \;=\; -\sum_{k\geq 2}\big(\varphi(k)-\mu(k)\big)\,W_k(\lambda)^2, \qquad W_k(\lambda) = \sum_{k|d}\frac{\lambda_d}{d}. \tag{8.3}$$

Every coefficient $\varphi(k)-\mu(k)$ is positive for $k \geq 2$, and every $W_k(\lambda)^2$ is a square, so the entire right-hand side is a sum of nonnegative terms with a minus sign in front. The centered form is therefore negative or zero for *every* choice of sieve weights: the matrix $B-M$ is negative semidefinite. Figure 8.2A is its spectrum, computed on the squarefree moduli coprime to 210 — one eigenvalue exactly zero, every other strictly below.

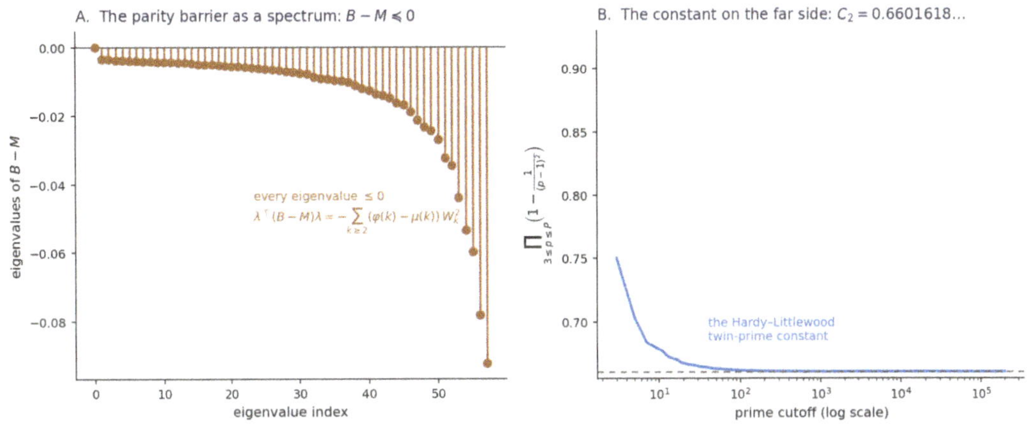

Figure 8.2: The twin wall, both sides. **A:** the spectrum of the centered matrix $B-M$, Eqs. (8.1)–(8.2), computed on squarefree moduli coprime to 210. Every eigenvalue is ≤ 0: the form is negative semidefinite, by the identity (8.3). This is the parity barrier as the sign of a quadratic form. **B:** on the far side, the Hardy–Littlewood twin-prime constant $C_2 = \prod_{p\geq 3}(1-1/(p-1)^2) = 0.6601618\ldots$, the number that would count the twin primes if there are infinitely many.

This negative-semidefiniteness is not a technicality; it is the parity barrier in finite dimensions. The barrier, in its classical form, is the fact that sieve methods cannot tell a prime from a product of two primes — cannot resolve the parity of the number of prime factors — and so cannot, on their own, force the count of primes (or twin primes) to be positive. Equation (8.3) is that obstruction made explicit and finite: the overlap form B never exceeds the smooth form M in the centered sense, so no choice of weights λ turns the sieve into a positive lower bound for twins. The best unconditional result, Chen's theorem, lives exactly at this edge: it gives infinitely many primes p for which $p+2$ is either a prime or a product of two primes, in each corridor — everything the method yields right up to the barrier, and not the final step across it, because that step is the one the parity barrier forbids.

8.5 The constant on the far side

Turn now to the far side of the wall and ask what number the spectral construction produces when pointed at the twins. Each of the two prime-side channels of the earlier chapters carries an energy — call them E_ψ and $E_{\psi^\dagger}$ — and the natural scalar the two-channel system forms is the product, $4\,E_\psi E_{\psi^\dagger}$. (The construction that produced these channels dressed this product as a relativistic mass; that dressing is a costume, and its own authors decline to attach it to any physical mass. The honest object is a product of two energies.) Follow that invariant up the primorials M_k, with p_k the largest prime dividing M_k, and it has a closed asymptotic form:

$$4\,E_\psi E_{\psi^\dagger}\,(M_k) \;\sim\; \frac{4\,C_2\,e^{-\gamma}\,S(\infty)}{\log p_k}, \tag{8.4}$$

where γ is the Euler–Mascheroni constant, $S(\infty)$ is a fixed number set by the windowing, and $C_2 = 0.6601618\ldots$ is the Hardy–Littlewood twin-prime constant of Figure 8.2B. The mechanism is Mertens' third theorem — the product $\prod_{p\le x}(1-1/p) \sim e^{-\gamma}/\log x$ — applied to the channel energies, with a leading-order cancellation that leaves C_2 standing.

The overall scale in (8.4) depends on the windowing convention and carries no meaning; what does not depend on any convention is which constant appears. Of all the numbers the construction could have produced for its most concrete, "physical-feeling" quantity, the one it produces is C_2 — the constant in the Hardy–Littlewood conjecture $\pi_2(x) \sim 2C_2\,x/(\log x)^2$, the number that would give the asymptotic count of twin primes if the conjecture holds. The spectral construction reaches, on the far side of the barrier, the exact constant that would weigh the twins. It reaches the number; it does not reach the theorem.

8.6 What this is and is not

What is established is two exact facts on opposite sides of the wall. On the near side, the centered identity (8.3) is elementary and proved, and the matrix $B - M$ is negative semidefinite — a finite, displayable form of the parity barrier, with Chen's theorem sitting at its edge. On the far side, the two-channel invariant has the closed asymptotic form (8.4), in which the Hardy–Littlewood constant C_2 provably appears by Mertens' theorem. Both are real mathematics; the figure computes both directly.

What is not established is the twin-prime conjecture, and nothing here moves it. The localization to fifteen corridors loses no difficulty; the negative-semidefiniteness is the obstruction, not a way around it; and the appearance of C_2 is the appearance of a number, not a proof of the infinitude it would measure. This is the honest shape of the second wall under the spectral lens. The barrier is the parity barrier, now visible as the sign of a small matrix one can write down; and the prize on the far side is a constant one can compute to as many digits as one likes, which is exactly not the same as proving there are infinitely many twins for it to count. Book One left the twins at their wall. This chapter shows the wall has a precise finite shadow and a precise number behind it, and stands exactly where it stood.

On the figure

Figure 8.2A is a direct computation: the matrices B and M of Eqs. (8.1)–(8.2) on the 58 squarefree moduli coprime to 210 below 260, and the eigenvalues of $B - M$, whose maximum is 0 to machine precision and whose remainder is negative — the negative-semidefiniteness of Eq. (8.3), checked rather than asserted. Panel B is the partial product $\prod_{3\leq p\leq P}(1 - 1/(p-1)^2)$, which converges to $C_2 = 0.6601618\ldots$ (at $P = 2 \times 10^5$ it reads 0.6601621).

The corridor wheel (Figure 8.1) is a direct enumeration: the fifteen residues r with both r and $r + 2$ coprime to 210, arranged by the mirror $r \leftrightarrow 210 - r$, with the primality of each member checked outright — which is what exposes the two broken slots.

Chapter 9

What the Zeros Don't Know

The third wall is the hardness of factoring, and it stands apart from the other two. The Riemann wall and the twin wall are places where the spectral approach reaches and then stops; the factoring wall is a place where it never points. The zeros of ζ encode the global distribution *of the primes — how many there are below a height, how they thin, how they fluctuate — and factoring asks a* local *question about one specific number: which two primes multiply to give it. These are different kinds of knowledge, and the first does not contain the second. Even a full proof of the Riemann Hypothesis, which would pin the distribution exactly, would not hand anyone an efficient way to factor. And the quantum that does break factoring — Shor's algorithm — works by a mechanism that has nothing to do with the quantum chaos of the zeros that this book has been about. There are two different "quantums" in this subject, and the one that threatens the codes is not the one that might prove the hypothesis. This chapter is short, because the honest thing to report about the third wall is mostly that the lens does not address it.*

9.1 Scope

This chapter is brief and assumes Book One's closing chapter, where the hardness of factoring was named as the third wall. It does not develop the theory of factoring or of public-key cryptography; it asks only what the spectral approach of this book has to say about it, and the answer is: very little, for a reason worth stating clearly. The point is the distinction between two kinds of knowledge about the primes, and the distinction between two unrelated uses of the word "quantum." Section 9.5 draws the line, and with it closes the book's walk along the three walls.

9.2 Two kinds of knowledge about primes

Everything in this book and the last has been about the distribution of the primes: their density on the wheel, their race across the residue classes, the smooth count of them below a height and the fluctuation around it, the zeros that govern that fluctuation. This is global knowledge. It answers questions of the form "how many primes are there with such-and-such a property," and it answers them in the aggregate, as statements about counts and averages and error terms.

Factoring asks something else entirely. Given a single large number N, known to be the product of two primes, find them. This is local knowledge — a question about the internal multiplicative structure of one specific integer, not about the population of primes around it. The two kinds of knowledge are genuinely different, and the gap between them is the reason the spectral approach falls silent here. Knowing the distribution of the primes with perfect precision tells you how many primes lie near $\sqrt{N}$; it does not tell you which of them divides N.

9.3 Why the zeros do not factor

The point can be made sharp through the explicit formula, the bridge of Book One. That formula relates a sum over the primes to a sum over the zeros, and what it computes is the count of primes (with prime powers, suitably weighted) below a height x. The zeros control the error in that count. So the information the zeros carry is exactly information about $\pi(x)$ — about how the primes are distributed — and the Riemann Hypothesis is the statement that this error is as small as it can be.

Suppose it were proved. Then the count of primes below any height would be known to within an error of order $\sqrt{x}\log x$, the smallest possible. This sharpens primality testing, tightens bounds on the gaps between primes, and improves a number of algorithms that depend on the distribution. It does nothing for factoring. The number field sieve, the best general factoring method known, is sub-exponential and makes no use of the zeros; no one has found a way to turn sharp knowledge of $\pi(x)$ into a fast way to split a specific N, and there is no reduction known in either direction between the Riemann Hypothesis and the hardness of factoring. The two problems are both about primes, and they are about different things primes do — spreading out, on the one hand, and multiplying together, on the other.

9.4 The other quantum

There is a real quantum threat to factoring, and it is worth naming precisely, because it is easy to conflate with the quantum spectrum this book has chased. Shor's algorithm factors a large number in polynomial time on a quantum computer. But it does not do so by finding a Riemann spectrum or diagonalizing a Hilbert–Pólya operator. It works by *period-finding*: to factor N, pick a random a and find the period of the sequence $a, a^2, a^3, \ldots$ modulo N — the multiplicative order of a — using a quantum Fourier transform on the cyclic group of exponents, after which an elementary greatest-common-divisor computation recovers the factors. The quantum work is a Fourier transform over the exponent register — the integers modulo a power of two of size about N^2 — from which the period is read off. It has nothing to do with the zeros of ζ, with quantum chaos, or with the statistics that filled the first half of this book.

So there are two quantums here, and keeping them apart is the whole content of this section. One is the quantum of *quantum chaos*: the idea, running through this book, that the zeros are the spectrum of some chaotic quantum system, and that proving the Riemann Hypothesis means finding that system. The other is the quantum of *quantum computation*: Shor's period-finding, which breaks the factoring-based codes and which is built from the quantum Fourier transform, not from any spectrum of the primes. The quantum that menaces the codes is not the quantum that might settle the hypothesis. They share a word and almost nothing else.

9.5 What this is and is not

What is established is a separation. The zeros of ζ carry information about the distribution of the primes, not about the factorization of any particular number; the Riemann Hypothesis, even proved, would not yield efficient factoring; and the quantum algorithm that does factor efficiently works by period-finding, unrelated to the spectral picture of the zeros. None of this is a theorem about factoring being hard — the hardness of factoring is itself unproven, a conjecture the codes rest on — but it is an honest account of why the instruments of this book do not touch it.

What this is not is a fourth approach to the third wall. The spectral lens reaches the Riemann wall and resolves it into a coupled family of walls; it reaches the twin wall and shows the parity barrier as the sign of a matrix, with the twin constant standing on the far side; and it does not reach the factoring wall at all, because that wall is built from a question the zeros were never about. Three walls, three different fates under the same lens — and that completes the walk. Book One ended by saying that the tour does not pass the three walls, and ends at them deliberately, because seeing exactly where a wall stands is worth more than a story that pretends to cross it. This book has tried to do the same thing one level deeper: to take the most physical-looking face of the subject, the zeros as a spectrum, and follow it honestly to each wall — to the operator that cannot be built, the resolution that cannot be had, the laws that do not survive being pushed, and the three obstacles that remain exactly where the first book left them. The spectrum is real, the statistics are real, the constants are real. The walls are real too, and they are still standing.

Chapter 10

Beyond the Circle

Every wheel in this book has secretly been a circle. There is a next curve, and it is not a figure of speech: it is the shape that appears when the rational integers give way to the Gaussian integers. We roll onto it to settle two questions at once—whether the walls were ever about the circle at all, and whether the one structure the whole book has leaned on, a world of positions facing a world of frequencies across the explicit formula, survives the change of stage. Both answers are the point of the book.

Scope

The wheel that has carried this book—the residues coprime to 210, turned and lifted and stressed—is, underneath its arithmetic, a circle. The roots of unity sit on it; the Dirichlet characters are its harmonics; the L-functions of Part IV are built from its division points. This closing chapter takes the circle seriously enough to ask what lies one step past it. The answer is a specific, named curve with a specific, named arithmetic behind it, and the correspondence is rigorous and more than a century old. We use it for two things. First, to test whether the three walls—the Riemann Hypothesis, the twin primes, factoring—are features of the particular wheel we chose or of the primes themselves, by changing the wheel as drastically as the mathematics allows and watching what stays. Second, to exhibit, one last time and in a new field, the single picture this whole book has been about: the primes living in one plane, their zeros living in another, and a single exact identity—the explicit formula—as the only bridge between them. Nothing here crosses a wall. Everything here is meant to show, cleanly, where the wall actually is.

10.1 The circle, made precise

Strip the wheel to its geometry and it is the unit circle with marked points. The marking is not arbitrary. The map

$$e : \mathbb{C}/\mathbb{Z} \longrightarrow \mathbb{C}^{\times}, \qquad z \longmapsto e^{2\pi i z},$$

is periodic with respect to the lattice $\mathbb{Z}$, and it sends the rational points of $\mathbb{C}/\mathbb{Z}$ to the roots of unity: $z = k/n$ goes to ζ_n^k, where $\zeta_n = e^{2\pi i/n}$. Dividing the circle into n equal arcs is exactly evaluating this one transcendental function at the n-division points of its lattice, and the values it returns generate the cyclotomic field $\mathbb{Q}(\zeta_n)$. The arithmetic of that field is as clean as a field gets: its Galois group is canonically

$$\mathrm{Gal}\big(\mathbb{Q}(\zeta_n)/\mathbb{Q}\big) \;\cong\; (\mathbb{Z}/n\mathbb{Z})^\times,$$

abelian, with the residue a mod n acting by $\zeta_n \mapsto \zeta_n^a$.

The theorem that makes the wheel more than bookkeeping is the one of Kronecker and Weber: every finite abelian extension of the rationals lies inside some cyclotomic field $\mathbb{Q}(\zeta_n)$. Put plainly, all of the simplest arithmetic over $\mathbb{Q}$—every way of extending the rationals with an abelian symmetry—is already written in the division points of the circle. The residues mod 210, the real characters and their conductor lattice, the family of Dirichlet L-functions whose zeros formed the family of walls: every one of them is the circle, divided. When we rolled the wheel we were stepping around the circle; when we spun it onto its dual face we were reading the circle's harmonics. The circle was the whole stage.

10.2 The next curve, made precise

In the 1820s Abel, following Gauss, asked what happens if you divide a different curve: the lemniscate, the figure-eight given in polar form by $r^2 = \cos 2\theta$. The circle's division was run by the exponential and its lattice $\mathbb{Z}$; the lemniscate's is run by a different transcendental function and a different lattice. Define the *lemniscate constant*

$$\varpi \;=\; 2\int_0^1 \frac{dt}{\sqrt{1-t^4}} \;=\; 2.6220575542\ldots,$$

the analogue of $\pi = 2\int_0^1 dt/\sqrt{1-t^2}$, and define the *lemniscate sine* $\mathrm{sl}(u)$ by inverting the arc-length integral, so that

$$u \;=\; \int_0^{\mathrm{sl}(u)} \frac{dt}{\sqrt{1-t^4}}, \qquad \mathrm{sl}'(u)^2 \;=\; 1-\mathrm{sl}(u)^4.$$

Like the sine, sl extends to a meromorphic function on the whole plane; unlike the sine, it is doubly periodic, an elliptic function of order two. Its period lattice is *square*: it is carried to itself by multiplication by i, and the function obeys the single identity that names the whole subject,

$$\mathrm{sl}(i\,u) \;=\; i\;\mathrm{sl}(u).$$

This is *complex multiplication.* The circle's symmetry was multiplication by roots of unity, a rotation; the lemniscate's symmetry includes a genuine multiplication by the imaginary unit, a quarter-turn of the lattice that the circle simply does not have. The curve standing behind sl is the elliptic curve

$$y^2 = x^3 - x, \qquad j\text{-invariant } 1728,$$

whose endomorphism ring is the full ring of Gaussian integers $\mathbb{Z}[i]$ and which carries the order-four automorphism $(x, y) \mapsto (-x, iy)$—one checks $(iy)^2 = -y^2 = (-x)^3 - (-x)$ directly. That automorphism *is* multiplication by i on the curve.

The historical fingerprint is unmistakable. Gauss had shown that a regular n-gon is constructible with ruler and compass when n is a power of two times distinct Fermat primes (Wantzel proved the converse); Abel showed that the lemniscate can be cut into n equal arcs by ruler and compass under the very same condition. The same Fermat primes governing two different curves is not a coincidence. The lemniscate's n-division values—the numbers $\mathrm{sl}(\varpi\beta/n)$ for Gaussian integers β—generate abelian extensions of $\mathbb{Q}(i)$, exactly as the roots of unity generate those of $\mathbb{Q}$. And the converse holds: every abelian extension of $\mathbb{Q}(i)$ is generated by such special values together with roots of unity. This is the imaginary-quadratic case of Hilbert's twelfth problem—Kronecker's *Jugendtraum*—solved by the theory of complex multiplication in the hands of Kronecker, Weber, Takagi, Hecke, and Deuring. The circle solves Hilbert's twelfth problem for $\mathbb{Q}$; the lemniscate solves it for $\mathbb{Q}(i)$. The wheel was the first case of a pattern. The lemniscate is the second (Figure 10.1).

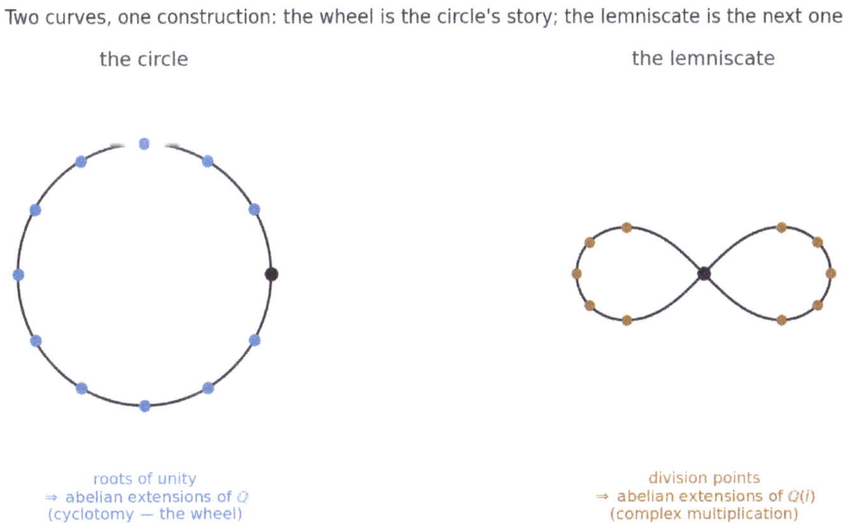

Figure 10.1: The circle and the lemniscate, divided. The circle's division points—the roots of unity—generate the abelian extensions of $\mathbb{Q}$; this is the wheel. The lemniscate's division points generate the abelian extensions of $\mathbb{Q}(i)$ by complex multiplication. The crossing node, in black, is the geometric home of the ramified prime 2.

10.3 How the primes map

Changing the base field from $\mathbb{Q}$ to $\mathbb{Q}(i)$ does something concrete to the primes themselves: each one is told to factor again, this time in the larger ring $\mathbb{Z}[i]$. That ring is Euclidean for the norm $N(a + bi) = a^2 + b^2$, hence a unique-factorization domain;

its units are the four numbers $\{1, i, -1, -i\}$, the quarter-turns. The rule by which a rational prime p factors is the exact analogue of how p splits in the cyclotomic world, and it is one of the oldest theorems in the subject, Fermat's on sums of two squares:

$$\begin{aligned} p \equiv 1 \pmod 4 &: \quad p = \pi\bar{\pi} \text{ \textit{splits} into two conjugate Gaussian primes, } p = a^2 + b^2, \\ p \equiv 3 \pmod 4 &: \quad p \text{ stays \textit{inert}, a Gaussian prime of norm } p^2, \\ p = 2 &: \quad 2 = -i\,(1+i)^2 \text{ \textit{ramifies}.} \end{aligned}$$

So $5 = (2+i)(2-i)$, $13 = (3+2i)(3-2i)$, $17 = (4+i)(4-i)$, while $3, 7, 11, 19$ remain prime where they stand, and 2 alone collapses onto a single repeated factor. This *is* the map from the rational primes to the Gaussian primes the chapter promised: it is the sum-of-two-squares decomposition, nothing more exotic.

What decides the three cases is a single arithmetic switch. The Galois group $\mathrm{Gal}(\mathbb{Q}(i)/\mathbb{Q})$ has two elements, the identity and complex conjugation, and the Frobenius element attached to an odd prime p is conjugation when p is inert and the identity when p splits. That switch is exactly the nontrivial character modulo 4,

$$\chi_4(p) = (-1)^{(p-1)/2} = \begin{cases} +1, & p \equiv 1 \pmod 4 \ \text{(split)}, \\ -1, & p \equiv 3 \pmod 4 \ \text{(inert)}, \end{cases}$$

with $p = 2$ the ramified exception where χ_4 is not defined. The same χ_4 assembles the whole story into one function. The Dedekind zeta function of $\mathbb{Q}(i)$, summed over the nonzero ideals of $\mathbb{Z}[i]$, has an Euler product over the Gaussian primes that the splitting law factors cleanly:

$$\zeta_{\mathbb{Q}(i)}(s) \;=\; \sum_{0 \neq I \subseteq \mathbb{Z}[i]} \frac{1}{N(I)^s} \;=\; \prod_{\pi} \left(1 - N(\pi)^{-s}\right)^{-1} \;=\; \zeta(s)\,L(s, \chi_4).$$

The right-hand side is the entire map, read as a single object: the factor $\zeta(s)$ is the rational backbone, and the factor $L(s, \chi_4)$ is precisely the splitting switch, the Frobenius, turned into a function of s. The Gaussian primes and the character χ_4 are two views of one thing.

10.4 Two planes, and the bridge between them

Now the crux of the book, in the cleanest field we have for it.

The Gaussian primes live in a *position plane*. They are points of the lattice $\mathbb{Z}[i]$ that you can name and place: $1+i$ at the centre, the split primes $2 \pm i$, $3 \pm 2i$ off the axes, the inert primes $3, 7, 11$ marching out along them. Multiplication by i spins this plane through a quarter turn and carries the picture to itself; complex conjugation $z \mapsto \bar{z}$ reflects it across the real axis, swapping the two conjugate factors π and $\bar{\pi}$ of every split prime. These are the roll and the barrel-roll of the wheel, now literal motions of a lattice in the plane.

The zeros of $\zeta_{\mathbb{Q}(i)} = \zeta \cdot L(\chi_4)$ live in a different plane entirely—a *spectral plane*, the strip of complex s, where under the Generalized Riemann Hypothesis they sit in a single line at $\mathrm{Re}(s) = \frac{1}{2}$. Because the function factors, its zeros are the union of two families, the zeros of ζ and the zeros of $L(\chi_4)$, interleaved on that line; there are more of them, and they are denser, than the circle's alone. This is the dual face: the same arithmetic, read as frequencies rather than as positions.

The two planes are not the same plane, and no amount of drawing puts a zero onto the lattice or a Gaussian prime onto the critical line. What connects them is one exact identity. Weil's explicit formula for $\zeta_{\mathbb{Q}(i)}$ states that a weighted sum over the Gaussian primes equals a weighted sum over the zeros, term for term, with only the archimedean factors and the pole at $s = 1$ to mediate. The primes on the left, the zeros on the right, an equals sign in between: that identity is the only legitimate bridge between the position plane and the spectral plane, and it is the same bridge—the same explicit formula read as a trace—that has stood at the centre of every chapter of this book. Figure 10.2 draws the two planes side by side, with the bridge named and nothing faked across it. If there is one picture to carry away, it is this one: the primes here, the zeros there, and between them a single conserved identity and not one inch of shortcut.

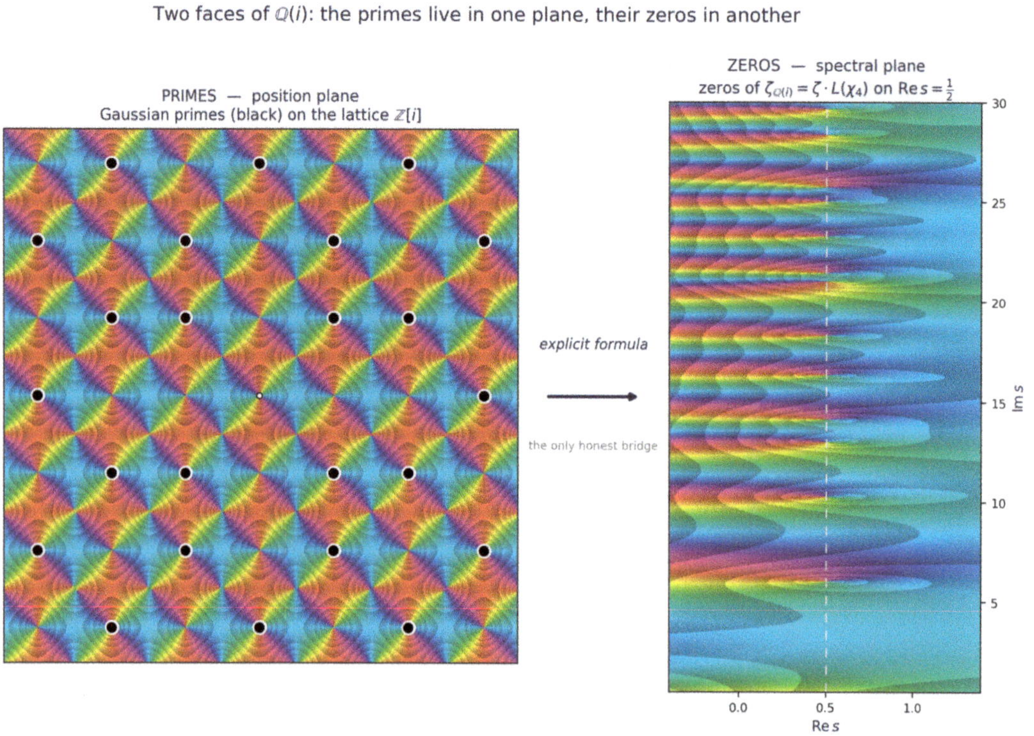

Figure 10.2: The crux of the book, in $\mathbb{Q}(i)$. *Left, the position plane:* the Gaussian primes (marked) sit among the integers on the lattice $\mathbb{Z}[i]$, drawn through the lemniscatic function whose poles are the lattice points; the quarter-turn $\times i$ is the spin, the dashed axis the conjugation mirror. *Right, the spectral plane:* a phase portrait of $\zeta_{\mathbb{Q}(i)} = \zeta \cdot L(\chi_4)$ over the critical strip, its zeros the dark points lined up on $\mathrm{Re}(s) = \frac{1}{2}$, the union of the zeta and $L(\chi_4)$ families. *Between them:* the explicit formula, the one exact identity equating a sum over the primes on the left with a sum over the zeros on the right. The zeros cannot be drawn on the lattice and the primes cannot be drawn on the line; the bridge is the only honest crossing. The two textures differ for a reason: the lattice on the left is periodic, a true crystal, so its colours repeat in tight identical cells, while $\zeta_{\mathbb{Q}(i)}$ on the right is aperiodic, its phase spreading in broad bands and its zeros forming not a lattice but a quasicrystal—ordered without ever repeating.

10.5 The walls travel

Here is the payoff of changing the stage. The move to the lemniscate is not a loss of structure—it is a gain, and a large one. Because the Gaussian world has complex multiplication, its L-functions are unusually well behaved: Hecke proved the analytic continuation and functional equation of the relevant Grössencharakter L-functions outright, decades before the general case was within reach, and the first genuine progress on the Birch and Swinnerton-Dyer conjecture was made in exactly this complex-multiplication setting, by Coates and Wiles, who proved that a CM elliptic curve with infinitely many rational points has L-value zero at $s = 1$. If any change of curve were going to soften the walls, this one—to the most structured alternative the subject offers—should have done it.

It does not. The Generalized Riemann Hypothesis for $\zeta_{\mathbb{Q}(i)} = \zeta \cdot L(\chi_4)$, and for the Hecke L-functions of $\mathbb{Q}(i)$ generally, is open, precisely as it is for the Dirichlet L-functions of the wheel; the extra structure that gives us their functional equations says nothing about where their zeros lie. The Gaussian-integer analogues of the twin-prime and gap questions—whether the split primes cluster, how far apart consecutive Gaussian primes can spread, whether one can even walk to infinity among them in steps of bounded length, the still-unsolved Gaussian moat—are open. Factoring a Gaussian integer is no easier than factoring a rational one; the same problem, in a richer ring, is the same problem. The three walls of Part IV reappear on the figure-eight, in the same places, with the same hardness, in full view of all the additional symmetry the new curve brings. The structure was real and it was generous, and it left the walls exactly where it found them.

What this is and is not

The correspondence in this chapter is established and celebrated mathematics. The exponential divides the circle and its values generate the abelian extensions of $\mathbb{Q}$; the lemniscate sine divides the figure-eight and its values generate the abelian extensions of $\mathbb{Q}(i)$; complex multiplication, $\mathrm{sl}(iu) = i\,\mathrm{sl}(u)$, makes the analogy exact; this is the content of Hilbert's twelfth problem in its solved cases. The splitting law is Fermat's theorem on sums of two squares, and the factorization $\zeta_{\mathbb{Q}(i)} = \zeta \cdot L(\chi_4)$ is its restatement as a single function. The explicit formula linking the Gaussian primes to the zeros of that function is exact. These are facts, and they are rigorous.

What this chapter is not is an attack on anything. Moving to the lemniscate does not weaken a single wall; it was never meant to. There is no lever in the extra symmetry, no shortcut hidden in the quarter-turn—if anything the lemniscate's lesson is the opposite of a shortcut. We changed the curve to the most natural successor mathematics offers, gained complex multiplication and everything that comes with it, and found the three walls standing where we left them. And the deeper observation, the one the whole book has been building toward, also survived the change without a scratch: there is a plane where the primes live and a different plane where their zeros live, and the only passage between the two is one exact identity. That structure is not a property of the circle, or of the lemniscate, or of any curve we choose to roll the primes along. It travels with the primes, onto every shape that will hold them—and so, it seems, do the walls.

On the figures

Figure 10.1 sets the two curves side by side: the unit circle with its roots of unity, and the lemniscate $r^2 = \cos 2\theta$ with division points along each lobe and the crossing node, the home of the ramified prime, in black. The division points there are schematic—

equal steps along each curve rather than the exact lemnatomic positions—and show the parallel construction, divide the curve and read off the field, not the computed values.

Figure 10.2 is the chapter's thesis in one frame. Its left panel is a phase portrait of the lemniscatic function over $\mathbb{Z}[i]$: hue encodes the phase, the bright lattice points are its poles, the Gaussian primes are marked among them, and the spin $\times i$ and conjugation mirror are drawn in. Its right panel is a phase portrait of $\zeta_{\mathbb{Q}(i)}(s)$ over the critical strip computed directly from $\zeta(s)\,L(s,\chi_4)$, with the zeros appearing as the dark phase singularities on the line $\mathrm{Re}(s) = \frac{1}{2}$. The arrow between the panels is the explicit formula. The two phase portraits are genuine computations of two different functions on two different planes; the figure places them together precisely to show that they cannot be overlaid—only bridged.

Coda: geometry that hits walls

A last caution to close on, and it is about a word. Three different objects in the first two books have carried the name Möbius, and they are not the same object. There is the Möbius *strip*, the one-sided band of topology. There is the Möbius *function* $\mu(n)$, the signed indicator of the squarefree integers that runs the supersymmetric primon gas of the companion volume. And there is the Möbius *transformation*, the conformal motion $z \mapsto (az + b)/(cz + d)$ that rotates the sphere. They share a man—August Möbius—and nothing else. To let the shared name suggest a shared mathematics would be the most refined version of the mistake these books have warned against throughout: not a coincidence of shape, like the Lorentz mass of Chapter 4, but a coincidence of *spelling*. We keep the three apart on purpose.

It is the third Möbius, the transformation, that genuinely moves the flat torus. The flat torus $(\cos u, \sin u, \cos v, \sin v)$ sits on the three-sphere as a product of two circles, every pair of them linked; it is the same Clifford torus on which the two-mode state of Chapter 4 lives. The conformal motions of the three-sphere—the Möbius transformations—carry it to the cyclides of Dupin, and a rotation that sweeps the torus through the point of projection is exactly the motion that turns it inside out, the doughnut everting through its own hole (Figure 10.3). That eversion is real geometry. It is also, one notes with the discipline of Chapter 4 in hand, an element of a group isomorphic to a Lorentz group—the conformal group always is—and that isomorphism is, once more, an invitation to a metaphor and not a derivation of a world.

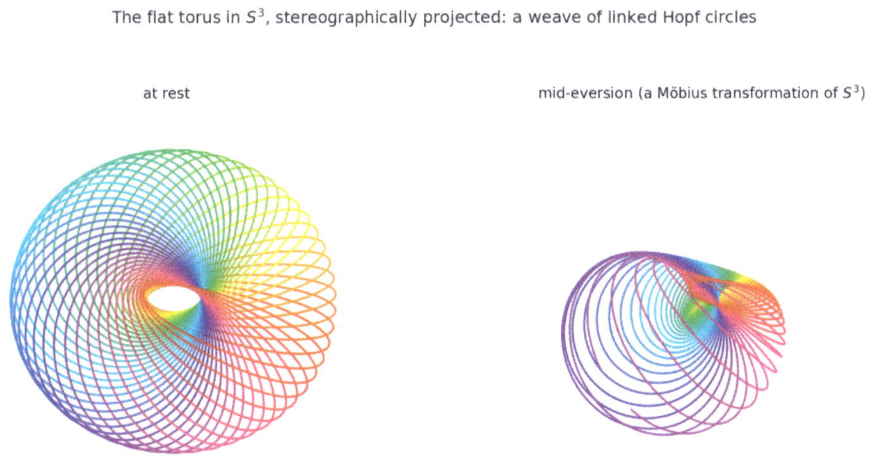

Figure 10.3: The flat torus $(\cos u, \sin u, \cos v, \sin v)$ of S^3, stereographically projected to ordinary space and drawn as a weave of its Hopf circles. Each circle winds once in both directions, and any two of them are linked. *Left:* at rest, the linked-circle torus, its central hole visible. *Right:* a rotation of S^3—a Möbius transformation—sweeps it toward the point of projection and turns it inside out, the image passing through the cyclides of Dupin. No rotation of the sphere changes which circles are linked, or where the wall is.

So take the dualities that have recurred across both books and ask which the geometry can actually carry. Two of them it carries cleanly. *Inside and outside* are the two solid tori into which the Clifford torus divides the three-sphere, exchanged by the eversion. *Position and spectrum* are the flat torus and its dual lattice, joined by Poisson summation, $\sum_{x\in\Lambda} f(x) = \sum_{\xi\in\Lambda^*} \hat{f}(\xi)$—a sum over a lattice equal to a sum over its dual, the explicit formula in its cradle. The third duality, *boson and fermion*, the geometry cannot carry, and it fails at a number. The natural supersymmetry of a torus is its de Rham grading, with Witten index equal to the Euler characteristic; and $\chi(T^2) = 0$, while the supersymmetric primon gas of the companion volume has Witten index 1. The invariant that defines the boson–fermion structure reads 0 on the torus and 1 in the gas. They are not the same structure, and no rotation of the sphere will make them one.

There is one further carrier, and it lives on a different surface from the flat torus—a curved one, reached by the same word through a different door. Restrict the Möbius transformations to integer coefficients and let them act not on the sphere, where they everted the torus, but on the hyperbolic upper half-plane; there they form the modular group, and the quotient is the modular surface, of constant negative curvature. On a flat surface the trace formula is just the Poisson summation of a moment ago. On this curved one it is Selberg's—an exact identity equating the spectrum of the hyperbolic Laplacian to the lengths of the closed geodesics, the surface's own primes—and it carries the one thing the Riemann case has never offered: the nontrivial zeros of the surface's zeta function, the Selberg zeta, *are* the eigenvalues of a self-adjoint operator, the Laplacian, and so they lie on their critical line for the most direct reason there is—the Hilbert–Pólya mechanism, not conjectured but working. It is realized in full, with a real operator and a real spectrum. But the geometry that does it is the hyperbolic modular surface, not the flat torus, and the zeta it does it for is Selberg's, not Riemann's. The same name, Möbius, has carried us to two stages—the flat torus it everts, the curved surface it tiles—and only one of them comes with an operator; and even that one is one zeta short of the wall.

That is the closing image of the first two books. The dualities are real and the geometry that carries them is beautiful; one of its carriers even crosses an analogue of the wall in plain sight, with an honest operator behind it. And Riemann's wall still stands—not because the geometry failed before reaching it, but because the wall was never a feature of the circle, or the torus, or the modular surface, or any shape onto which we can roll the primes. It belongs to the primes. Geometry that beautiful, running that far, and stopping exactly there: it is the truest picture the subject can offer of where the difficulty lives, and a genuinely beautiful one.

Exercises

The exercises are collected here, by chapter, rather than interrupting the text. Each exercise is tagged by difficulty ([routine], [moderate], [hard]) or by kind ([conceptual], [discussion]). The *two-part test* referred to throughout is that of Chapter 2: a candidate Riemann operator must (i) have the nontrivial zeros as its spectrum and (ii) be defined independently of the zeros.

Chapter 1 The Symmetry of a Family

1. [routine] Describe qualitatively the behaviour of the one-level density near the central point for a unitary family, a symplectic family, and an orthogonal family. Which is suppressed at the centre, which enhanced, and which flat?

2. [moderate] Distinguish *bulk universality* (pair correlation following the unitary law for every family) from the *central-point statistics* that separate the symmetry types. Why can the bulk statistics alone not reveal a family's symmetry type?

3. [moderate] Over function fields the symmetry type of a family is a theorem; over $\mathbb{Q}$ it is conjectural, supported by one-level density results valid only for a restricted range of test functions. What does the function-field setting supply that $\mathbb{Q}$ does not?

4. [discussion] Explain why identifying a family's symmetry type does not, by itself, establish that any individual L-function has its zeros on the critical line.

5. [moderate] Each of the seven real L-functions of the wheel, measured alone, has a level-spacing ratio in the unitary range ($\langle \tilde{r} \rangle \approx 0.62$); pooling all seven sets of zeros into a single list drops the ratio to the uncorrelated value (≈ 0.39, Figure 1.2). Explain why superimposing several rigid, level-repelling spectra can erase the repulsion this completely only when the spectra are mutually independent, and say what a *surviving* repulsion in the merged list would have shown instead.

Chapter 2 What Would Count as an Operator

1. [routine] *The primon spectrum.* The primon gas has one mode per prime p with single-quantum energy $\log p$ and occupation numbers $a_p \in \{0, 1, 2, \dots\}$.

(a) Show the energy $E = \sum_p a_p \log p$ of a state equals $\log n$, where $n = \prod_p p^{a_p}$.
(b) Using unique factorization, show the energy spectrum is exactly $\{\log n : n = 1, 2, 3, \dots\}$.
(c) Conclude the partition function is $\sum_{n\geq 1} n^{-s} = \zeta(s)$ for $\mathrm{Re}(s) > 1$.

2. [moderate] *Counting the levels.* Let $N_{\mathrm{pr}}(E)$ count primon levels $\leq E$.

   (a) Show $N_{\mathrm{pr}}(E) = \lfloor e^E \rfloor \sim e^E$.
   (b) Evaluate at $E = \gamma_1 = 14.1347$, the first zero's height.
   (c) With $N(T) \approx \frac{T}{2\pi} \log \frac{T}{2\pi} - \frac{T}{2\pi} + \frac{7}{8}$, evaluate at $T = \gamma_1$ and compare. By how many orders of magnitude do the counts differ?

3. [routine] State the two-part test in your own words. For each candidate say which part(s) it passes and fails: (a) a diagonal operator whose eigenvalues are declared to be the γ_n; (b) the primon Hamiltonian; (c) a kernel of delta spikes placed at the γ_n.

4. [moderate] Explain precisely why building an operator whose eigenvalues are by definition the zeros tells us nothing about RH. What property must such a construction have to bear on RH at all?

5. [moderate] The primon gas reproduces $\zeta(s)$ as its partition function yet its spectrum is $\{\log n\}$, not the zeros. Reconcile these, distinguishing the *values* of ζ (its Dirichlet series) from the *location of its zeros.*

6. [hard] (a) Show the primon density of states near E grows like e^E. (b) Show the density of zeros near T grows like $\frac{1}{2\pi} \log \frac{T}{2\pi}$. (c) Explain why this mismatch alone proves the primon gas is not a Riemann operator, with no zero located.

Chapter 3 Primes as Orbits

1. [moderate] The Berry–Keating model recovers the smooth counting term from a phase-space volume. Verify it gives $N(T) \approx \frac{T}{2\pi} \log \frac{T}{2\pi} - \frac{T}{2\pi}$, and evaluate at $T = 100$: the leading volume gives ≈ 28, while the true count is 29 (the gap of about one being the $\frac{7}{8}$ term plus fluctuations).

2. [moderate] In the trace-formula reading the primes are periodic orbits of period $\log p$ (and p^k contributes period $k \log p$). Explain how the explicit formula takes the shape "sum over zeros = smooth term + sum over orbits."

3. [routine] Explain why xp passes the independence half of the two-part test. Which candidate of Chapter 2 fails independence, and how does xp avoid that failure?

4. [discussion] Berry–Keating recovers the *average* (smooth count) exactly but not the fluctuations or individual zeros, the obstruction being that the required dynamics is chaotic and no concrete system realizing it is known. Explain what "average but not fluctuations" means here.

5. [hard] Both the primon gas (Chapter 2) and xp involve the period $\log p$. Contrast why the primon gas fails on spectrum while xp at least captures the average correctly.

Chapter 4 An Algebra in the Primes

1. [moderate] The prime bilinears close into a Lie algebra, $\mathfrak{sp}(4, \mathbb{R}) \cong \mathfrak{so}(2, 3)$. Explain what "close into an algebra" means in terms of commutators.

2. [moderate] The oscillator (metaplectic) representation splits into two irreducible singletons, the Rac ($\Delta = \frac{1}{2}$) and the Di ($\Delta = 1$). Relate this split to the even/odd sectors.

3. [routine] Despite the richer algebra, the spectrum is still $\{\log n\}$. Why does possessing an algebra not, by itself, repair the spectrum half of the two-part test?

4. [moderate] State Connes's criterion at the headline level (RH equivalent to a certain positivity). Explain why an equivalence "RH $\Leftrightarrow X$" is not a proof of RH.

5. [discussion] de Branges's approach has a known obstruction (Conrey–Li). What does it mean for an approach to be "not obviously wrong but with a located gap," and why is that the normal state of affairs near this wall?

6. [routine] With $J^0 = E_+ + E_-$ and $J^z = E_+ - E_-$, show $(J^0)^2 - (J^z)^2 = 4E_+E_-$, so the "dispersion relation" $E^2 = m^2 + p^2$ (with $E = J^0$, $p = J^z$, $m^2 = 4E_+E_-$) is an algebraic identity. Explain why its holding "to machine precision" is not evidence that the primes encode relativistic kinematics.

Chapter 5 Hearing the Zeros

1. [moderate] Distinguish *detection* from *resolution*. The prime-side coherence is significant ($p \approx 5 \times 10^{-5}$) within one mean spacing of a zero yet cannot resolve individual zeros. What is the difference?

2. [moderate] The resolution limit (aperture) is set by log(largest prime used). Explain the analogy to a telescope aperture or a Fourier window: why does the largest prime fix the resolution?

3. [conceptual] Four super-resolution methods of different mechanism all fail below the aperture. Why is it significant that unrelated methods hit the *same* limit?

4. [hard] If primes up to X are used, the aperture is $\sim \log X$. Discuss how slowly the resolvable spacing improves as X grows, and what this implies for ever resolving individual zeros from the primes.

5. [discussion] Why is "detect but not resolve" the honest analog, in the measurement setting, of the operator dream's failure in the construction setting?

Chapter 6 Laws That Died

1. [routine] The seven real characters mod 210 correspond to the non-empty subsets of $\{3, 5, 7\}$. Fill in the truth table and verify the rule "sign $= x_5 \oplus x_7$" across all seven.

2. [moderate] The rule is clean to $X = 10^9$ on the wheel of 210 but fails on the wheel of 2310. Explain the two independent axes of stress (cutoff X, modulus M) and why surviving one does not imply surviving the other.

3. [conceptual] The same split rates 100.0th percentile under one observable and 43.7th under another at modulus 2310. What does this say about the "extremality" of the split?

4. [moderate] The archimedean explanation was rejected for being three orders of magnitude too small. Why is checking the *magnitude* of a proposed effect a basic and decisive test?

5. [discussion] Why is keeping a numbered list of falsified hypotheses the practice that separates this work from numerology?

6. [hard] A pattern holds for all primes up to 10^9 on one wheel. List the tests it must still pass before it deserves to be called a candidate law, and say what each test controls for.

Chapter 7 A Family of Walls

1. [conceptual] Once one works on the wheel, why is the relevant obstacle the *Generalized* Riemann Hypothesis (RH for every Dirichlet L-function) rather than RH for ζ alone?

2. [moderate] The residue $209 = 11 \cdot 19$ dominates the cancellation because it is built from the two characters most anti-correlated with ζ. Trace the chain: residue $\to$ characters $\to$ L-function zeros.

3. [conceptual] At modulus 210 the internal/external split is clean; at 2310 it dissolves, and one character even flips sign. What does this show about "which L-functions oppose ζ"?

4. [moderate] The explicit formula predicts the principal-character piece (about three-quarters of the observable at modulus 210) to roughly a percent, but not the rest. Why does the boundary between predictable and not coincide with the boundary between ζ and the other L-functions?

5. [discussion] "The spectral lens resolves one wall into a coupled family of walls." Explain why this is a larger obstacle than RH alone, not a smaller one.

Chapter 8 The Barrier and the Constant

1. [routine] Verify the two Möbius identities $\mathbf{1}_{\gcd(u,v)=1} = \sum_{k|\gcd} \mu(k)$ and $\gcd(u,v) = \sum_{k|\gcd} \varphi(k)$, from $\sum_{d|n} \mu(d) = \mathbf{1}_{n=1}$ and $\sum_{d|n} \varphi(d) = n$.
2. [moderate] On the squarefree moduli $\{1, 11, 13, 143\}$ coprime to 210, form

$$B(d_1, d_2) = \mathbf{1}_{\text{gcd}=1}/(d_1 d_2), \qquad M(d_1, d_2) = 1/\operatorname{lcm}(d_1, d_2),$$

   and check the eigenvalues of $B - M$ are ≤ 0.
3. [moderate] From the centered identity $\lambda^\top (B - M)\lambda = -\sum_{k\geq 2}(\varphi(k) - \mu(k))W_k^2$, deduce that $B - M$ is negative semidefinite. Why is $\varphi(k) - \mu(k) \geq 0$ for every $k \geq 2$?
4. [conceptual] Explain how the negative-semidefiniteness is a finite-dimensional form of the parity barrier, and where Chen's theorem sits relative to it.
5. [routine] Compute the partial product $\prod_{3\leq p\leq P}(1 - \frac{1}{(p-1)^2})$ for growing P and observe convergence to $C_2 = 0.6601618\ldots$
6. [discussion] The construction reaches C_2 but not the twin conjecture. Why is computing the constant not the same as proving there are infinitely many twins?

Chapter 9 What the Zeros Don't Know

1. [conceptual] Distinguish the two kinds of knowledge: the global distribution of the primes (what the zeros encode) and the local factorization of a specific N (what factoring needs).
2. [conceptual] Explain why a proof of RH would sharpen estimates of $\pi(x)$ and speed some primality tests, yet would not yield an efficient factoring algorithm.
3. [routine] Outline Shor's algorithm as period-finding (the multiplicative order of a random a modulo N) plus a greatest-common-divisor step, and identify where the quantum Fourier transform does the work.
4. [conceptual] Distinguish the "two quantums": quantum chaos (the zeros as a spectrum) and quantum computation (Shor). Why is conflating them an error?
5. [discussion] Of the three walls, why is factoring's relationship to the spectral approach different *in kind* from that of RH and the twins?

Chapter 10 Beyond the Circle

1. [routine] Verify the splitting of rational primes in the Gaussian integers: using Fermat's two-square theorem, show $p \equiv 1 \pmod 4$ factors as $(a+bi)(a-bi)$ (it splits), that $p \equiv 3 \pmod 4$ stays prime, and that $2 = -i(1+i)^2$ ramifies.

2. [moderate] Show $(x, y) \mapsto (-x, iy)$ maps $y^2 = x^3 - x$ to itself, has order four, and squares to $(x, y) \mapsto (x, -y)$. Explain why it realizes multiplication by i.

3. [conceptual] State the analogy precisely: roots of unity are to $\mathbb{Q}$ as the lemniscate's division points are to $\mathbb{Q}(i)$. Which theorem plays the role of Kronecker–Weber here, and of which problem is it a solved case?

4. [discussion] The lemniscate brings complex multiplication — more structure than the circle — yet the three walls persist over $\mathbb{Q}(i)$. Explain why "more structure" did not become "weaker walls," and what that says about where the difficulty resides.

5. [conceptual] The Gaussian primes are points of the lattice $\mathbb{Z}[i]$; the zeros of $\zeta_{\mathbb{Q}(i)} = \zeta \cdot L(\chi_4)$ are points of the critical strip. Explain why one cannot honestly plot the zeros on the lattice (or the primes on the line), and identify the single exact identity that legitimately relates the two. Why is its existence *not* a route across any wall?

Solutions to the Exercises

These are answer sketches, not full write-ups: enough to grade against and to run a discussion from. Exercise numbers match the *Exercises* section at the back of the book. The *two-part test* is that of Chapter 2 — a candidate Riemann operator must (i) have the zeros as its spectrum and (ii) be defined independently of them.

Chapter 1 The Symmetry of a Family

1. **Unitary:** flat at the centre ($W_U(x) = 1$) — no excess or deficit of low zeros. **Symplectic:** *suppressed* at the centre ($W(x) = 1 - \frac{\sin 2\pi x}{2\pi x} \to 0$ as $x \to 0$): zeros are repelled from the central point. **Orthogonal:** *enhanced* at the centre (even orthogonal, $W(x) = 1 + \frac{\sin 2\pi x}{2\pi x} \to 2$; the odd case carries a forced zero at the centre). So symplectic suppressed, orthogonal enhanced, unitary flat.

2. Pair correlation in the bulk follows the universal sine-kernel (unitary) law for *every* family, so it is identical across symmetry types and cannot tell them apart. The type shows only in the lowest zeros near the central point — the one-level density there — which differs by type. Bulk statistics are therefore blind to the distinction.

3. Over function fields the zeros are eigenvalues of Frobenius on cohomology: RH is a theorem (Weil), and the symmetry type follows from the monodromy group via Deligne's equidistribution (Katz–Sarnak). The setting supplies a genuine operator with a provable spectrum and a group-theoretic source for the type. Over $\mathbb{Q}$ neither exists; the one-level density is known only for test functions of restricted support (Özlük–Snyder for the quadratic family, Iwaniec–Luo–Sarnak more generally), so the type is inferred in that window, not proved.

4. The symmetry type is an *average* over a whole family; an average can hold while individual members deviate, and it is a statement about the statistics of zeros, presupposing their location rather than fixing it. So it constrains no single L-function's zeros.

5. Each L-function alone is a level-repelling (unitary) spectrum: its neighbouring zeros avoid one another, so very short gaps are rare. Merge several *independent* such sequences and they interleave blindly — neither knows where the other's

zeros sit — so a zero of one family can fall arbitrarily close to a zero of another. The short gaps that repulsion forbids *within* a family reappear *between* families, and the merged spacings relax to those of uncorrelated points ($\langle\tilde{r}\rangle \to 0.386$). Independence is the one arrangement that lands the merge exactly on Poisson; dependence shows as a departure in either direction. If the families' zeros tended to coincide or correlate, the pooled list would bunch into near-coincidences and the ratio would fall *below* 0.386; if instead the seven were jointly rigid — slices of one repelling spectrum — their mutual repulsion would survive the merge and hold the ratio *above* 0.386. The observed collapse to ≈ 0.39 is thus the statistical independence of the seven (Rudnick–Sarnak) made visible; a departure either way would have exposed a dependence among them.

Chapter 2 What Would Count as an Operator

1. (a) $\sum_p a_p \log p = \log \prod_p p^{a_p} = \log n$. (b) Unique factorization is the bijection between occupation vectors (a_p) and integers $n \geq 1$. (c) $\sum_{n\geq 1} n^{-s} = \zeta(s)$.

2. (a) $\log n \leq E \Leftrightarrow n \leq e^E$, so the count is $\lfloor e^E \rfloor \sim e^E$. (b) $\lfloor e^{14.1347} \rfloor \approx 1.38 \times 10^6$. (c) The smooth term at $T = \gamma_1$ is ≈ 0.45, so $N(\gamma_1)$ is essentially the first zero, ≈ 1; the counts differ by about *six orders of magnitude.*

3. (a) spectrum yes, independence no (circular); (b) independence yes, spectrum no; (c) spectrum yes, independence no.

4. The eigenvalues are the *input*, not a consequence, so nothing is predicted. To bear on RH the spectrum must be *forced* by a structure defined without reference to the zeros — i.e. the construction must pass independence.

5. The partition function encodes ζ as a Dirichlet series; the spectrum $\{\log n\}$ records the series' *exponents* (equivalently the Euler product). The zeros are an analytic-continuation feature, not visible as energy levels. Values and zeros are different features of the same function.

6. (a) $N_{\text{pr}}(E) \sim e^E \Rightarrow dN/dE \sim e^E$. (b) $\frac{d}{dT}\left[\frac{T}{2\pi}\log\frac{T}{2\pi} - \frac{T}{2\pi}\right] = \frac{1}{2\pi}\log\frac{T}{2\pi}$. (c) The counting function is an operator invariant; an exponential density of states cannot equal a logarithmic one, so the primon Hamiltonian is not a Riemann operator — no zero located.

Pitfall: *the recurring error is reading "reproduces ζ" (Ex. 1c) as "has the zeros as its spectrum." Ex. 5 exists to break that.*

Chapter 3 Primes as Orbits

1. The semiclassical state count for $H = xp$ gives $N(T) \approx \frac{T}{2\pi}\log\frac{T}{2\pi} - \frac{T}{2\pi}$. At $T = 100$: $15.9155 \times 2.7673 - 15.9155 \approx 28.1$ ($+\frac{7}{8} \approx 29.0$), matching the 29 actual zeros below 100.

2. The explicit formula equates a sum over zeros to a smooth (archimedean) term plus a sum over prime powers weighted by $\log p$ at "time" $k \log p$ — the shape of a trace formula (Gutzwiller): spectral sum = Weyl term + sum over periodic orbits, with p a primitive orbit of period $\log p$ and p^k its k-fold traversal.

3. xp is a classical Hamiltonian specified with no reference to the zeros, so its spectrum is not put in by hand: it passes independence. The Chapter-2 candidate that *fails* independence is the kernel assembled from the ordinates themselves (the delta/cosine kernel), which is circular. (xp shares with the primon gas that both are zero-independent; the primon gas's failure is on spectrum, not independence.)

4. "Average" is the smooth counting function / mean density of zeros, which Berry–Keating reproduces exactly. "Fluctuations" are the deviations of individual zeros from that mean; they require the detailed chaotic dynamics, and no concrete system realizing it is known, so the individual zeros are out of reach.

5. Both carry the period $\log p$, but the primon gas's spectrum is explicitly $\{\log n\}$, with exponentially growing density — wrong by construction. xp's phase-space count reproduces the correct (logarithmic) mean density, so it gets the average right and only the fluctuations are unknown. One has an explicit wrong spectrum; the other has the right smooth spectrum and an open fluctuation spectrum.

Chapter 4 An Algebra in the Primes

1. The commutators of the prime-bilinear operators are again linear combinations of the same operators — the set is closed under $[\cdot, \cdot]$ — so they span a Lie algebra, namely $\mathfrak{sp}(4, \mathbb{R}) \cong \mathfrak{so}(2, 3)$.

2. The oscillator representation acts on a Fock space graded by the parity of the number of quanta; the even and odd subspaces are each irreducible — the even one is the Rac singleton ($\Delta = \frac{1}{2}$), the odd one the Di singleton ($\Delta = 1$). The even/odd grading *is* the Rac/Di split.

3. The spectrum belongs to the chosen operator, not to the algebra around it. Embedding that operator in a richer algebra leaves its eigenvalues unchanged, so the spectrum half still reads $\{\log n\}$, not the zeros.

4. Connes: RH (for the relevant L-functions) is equivalent to the positivity of a certain trace (the Weil distribution). An equivalence relocates the difficulty — proving the positivity is exactly as hard as RH — so it is a reformulation, not a proof.

5. de Branges reduces RH to positivity/structure conditions on spaces of entire functions; Conrey–Li located a specific obstruction showing those conditions cannot hold in the required form. "Not refuted wholesale, but with a precise step

known to fail" is the normal state near this wall: approaches reach a definite, identifiable stopping point.

6. $(E_+ + E_-)^2 - (E_+ - E_-)^2 = 4E_+E_-$ by difference of squares, so $E^2 - p^2 = m^2$ holds for *all* values — it is an identity, not a measurement. Machine-precision agreement therefore carries no empirical content; the Lorentzian look comes entirely from contracting with $\eta = \text{diag}(+, -, -, -)$, a choice put in by hand. A satisfied identity is not a discovered law of the primes.

Chapter 5 Hearing the Zeros

1. **Detection** establishes that a signal correlated with the zeros is present (significant against a null). **Resolution** separates and locates individual zeros. One can detect that the primes track the zeros near a height without pinning the individual ordinates.

2. The primes act like baselines in an aperture of length log(largest prime), and Fourier/Rayleigh resolution scales inversely with that aperture. So the largest prime sets the finest resolvable spacing, as a telescope's diameter sets its angular resolution.

3. Different mechanisms hitting the same limit means the limit is in the information available (the finite aperture), not in any one algorithm. A method-specific failure could be engineered around; a shared wall is the signature of an information-theoretic bound.

4. With aperture $A = \log X$ the resolvable spacing scales like $1/\log X$; halving it needs $\log X$ doubled, i.e. X squared. Resolution improves only logarithmically in X, so resolving individual zeros would take exponentially more primes — detection is feasible, resolution is not.

5. In the construction setting the dream fails because no operator with the right spectrum can be built non-circularly; in the measurement setting the analog is that the primes carry detectable but insufficient information about the zeros. Both: the link is real, but stops short of pinning the zeros down.

Chapter 6 Laws That Died

1. $\{3\}(0,0)$ →U; $\{5\}(1,0)$ →O; $\{7\}(0,1)$ →O; $\{15\}(1,0)$ →O; $\{21\}(0,1)$ →O; $\{35\}(1,1)$ →U; $\{105\}(1,1)$ →U. OVER $= \{5, 7, 15, 21\}$, UNDER $= \{3, 35, 105\}$; the prime 3 never affects the sign.

2. The cutoff X and the modulus M are independent axes. Large X shows a pattern is not an artifact of small primes; a modulus lift shows it is not an artifact of the particular wheel. Passing one says nothing about the other — XOR passed X to 10^9 and failed the lift to 2310.

3. A property that flips from 100th percentile to below the random average depending on the observable is a property of the *observable*, not the split. The "extremality" is therefore observable-dependent, not intrinsic.

4. A proposed effect must be at least the size of what it explains. The archimedean piece is three orders of magnitude too small, so whatever its sign or structure it cannot close the gap — a magnitude check disqualifies it at once.

5. Recording falsified hypotheses in order forces the line between what survived testing and what merely appeared; it blocks quiet retention of dead patterns and the cherry-picking that turns coincidence into "law." That discipline is exactly what numerology lacks.

6. It should at least: survive a modulus lift (wheel-specific coincidence), survive a change of observable (measurement-dependence), beat a null/random ensemble (chance), and have any proposed mechanism checked for magnitude (size). Each guards a failure mode seen in the chapter.

Chapter 7 A Family of Walls

1. Every wheel observable is a prime sum sorted by residue, governed through the explicit formula by the zeros of *all* the wheel's Dirichlet L-functions, not just ζ. Controlling it needs RH for the whole family — GRH — strictly stronger than RH for ζ.

2. $209 = 11 \cdot 19$ dominates the cancellation; it is built from the conductor-11 and conductor-19 characters; those carry the L-functions most anti-correlated with ζ ($-0.82, -0.77$), i.e. whose zeros push hardest against ζ's. Residue $\to$ characters $\to$ L-function zeros.

3. At mod 210 internal characters move with ζ, external against; at mod 2310 the split dissolves and χ_{-19} flips sign. So "which L-functions oppose ζ" is not fixed — it depends on which primes divide the wheel.

4. The explicit formula can sum ζ's zeros, so it predicts the principal piece (built from ζ) to $\sim$1%. The other pieces are built from the non-principal L-functions, whose zeros it lacks. So the predictable/unpredictable boundary is exactly ζ vs the family — RH vs GRH.

5. RH concerns one function's zeros; the signal depends on the joint placement of an entire family's zeros, with couplings that reorganize per wheel, so there is no fixed reduction to one problem. The lens reveals a coupled, wheel-dependent family of obstacles — larger than the single wall, not a way past it.

Chapter 8 The Barrier and the Constant

1. From $\sum_{d|n}\mu(d) = \mathbf{1}_{n=1}$ with $n = \gcd(u,v)$, and noting the common divisors of u, v are the divisors of gcd, get $\mathbf{1}_{\gcd=1} = \sum_{k|\gcd}\mu(k)$. From $\sum_{d|n}\varphi(d) = n$ with $n = \gcd$, get $\gcd = \sum_{k|\gcd}\varphi(k)$.

2. On $\{1, 11, 13, 143\}$ the $d = 1$ row and column of $B - M$ vanish; the block on $\{11, 13, 143\}$ has diagonal $(-\frac{1}{11}, -\frac{1}{13}, -\frac{1}{143})$ with small off-diagonals, and its eigenvalues are all ≤ 0. So $B - M$ has one zero eigenvalue (the $d = 1$ direction) and three negative — negative semidefinite.

3. The identity writes $\lambda^\top(B - M)\lambda$ as minus a sum of $(\varphi(k) - \mu(k))$ times squares. Since $\varphi(k) \geq 1$ and $\mu(k) \leq 1$, each coefficient $\varphi(k) - \mu(k) \geq 0$ (indeed > 0 for $k \geq 2$), so the form is ≤ 0 for all λ: $B - M \preceq 0$.

4. Negative-semidefiniteness means no weights λ make the centered form positive — exactly the parity barrier (sieves cannot force a positive lower bound for twins) as the sign of a finite matrix. Chen's theorem sits at the edge: it gives everything up to the barrier (infinitely many p with $p+2$ prime or a product of two primes), not the step across.

5. The partial products $\prod_{3\leq p\leq P}(1-1/(p-1)^2)$ decrease toward $C_2 = 0.6601618\ldots$, agreeing to several digits well before P reaches the thousands.

6. C_2 is the constant in the Hardy–Littlewood asymptotic; computing it is evaluating a convergent product and presupposes nothing about whether the asymptotic — hence the infinitude of twins — holds. Reaching the number that would count the twins is not proving there are infinitely many to count.

Pitfall: *in Ex. 3, students sometimes try to show* $\varphi(k) - \mu(k) > 0$ *case-by-case; the one-line bound* $\varphi \geq 1 \geq \mu$ *suffices.*

Chapter 9 What the Zeros Don't Know

1. The zeros encode *global* information — counts and fluctuations of the primes in aggregate ($\pi(x)$ and its error). Factoring needs *local* information — the prime decomposition of one N. Knowing the distribution near $\sqrt{N}$ does not say which prime divides N.

2. RH sharpens the error in $\pi(x)$ (to $\sim\sqrt{x}\log x$), tightening prime-counting, gap bounds, and some primality tests — none of which factor a given N. The number field sieve uses no zero data, and no reduction from RH to fast factoring is known.

3. Pick random a coprime to N; find the multiplicative order r of a mod N (the period of a^x mod N) by a quantum Fourier transform over the exponents; if r is even and $a^{r/2} \not\equiv -1$, then $\gcd(a^{r/2} \pm 1, N)$ is a nontrivial factor. The speedup lives entirely in the QFT period-finding; the rest is classical gcd.

4. **Quantum chaos:** the conjecture that the zeros are the spectrum of a chaotic quantum system (Hilbert–Pólya / Berry–Keating). **Quantum computation:** Shor's period-finding via the QFT on the exponent register. Same word, different mathematics and different objects; conflating them wrongly suggests the device that threatens RSA bears on RH.

5. For RH and the twins the lens reaches the wall and stops there (it even reaches C_2). For factoring it does not point at the wall at all: the zeros are about distribution, factoring about local structure — so the spectral approach is not a stalled attempt but an irrelevant one. A different kind of relationship.

Chapter 10 Beyond the Circle

1. $p \equiv 1 \pmod 4 \Rightarrow p = a^2 + b^2$ (Fermat) $= (a + bi)(a - bi)$, two non-associate Gaussian primes — p splits. $p \equiv 3 \pmod 4$ is not a sum of two squares and stays prime (inert), of norm p^2. And $2 = (1+i)(1-i) = -i(1+i)^2$ up to a unit — a square of one prime, hence ramified.

2. $(-x)^3 - (-x) = -(x^3 - x) = -y^2 = (iy)^2$, so the image lies on the curve. Twice gives $(x, i^2y) = (x, -y)$ (the $[-1]$ map); four times the identity — order four. Squaring to $[-1]$ is the relation $i^2 = -1$, so the automorphism is multiplication by i.

3. Dividing the curve yields generators of the field's abelian extensions. For $\mathbb{Q}(i)$ the role of Kronecker–Weber is played by the main theorem of complex multiplication (the lemnatomic analogue); it is the solved case of Hilbert's twelfth problem for the imaginary quadratic field $\mathbb{Q}(i)$.

4. CM delivers analytic continuation and functional equations for the Hecke L-functions and the first partial Birch–Swinnerton-Dyer results (Coates–Wiles), so the Gaussian world is in many ways better understood. Yet GRH for those L-functions, the Gaussian twin and gap problems, and factoring all remain open. The extra structure improved what can be proved *around* the walls without moving them: the obstruction lies in the placement of the zeros, the parity of sieves, and the locality of factorization — which the change of base field reproduces rather than removes. The difficulty is a property of the primes, not of the curve.

5. In position you can name a Gaussian prime $(1 + i,\ 2 + i)$; the zeros live in the s-plane on $\operatorname{Re} s = \frac{1}{2}$ — different coordinates, different objects, so plotting one set on the other's axes asserts a coincidence that does not exist. The honest link is Weil's explicit formula, an exact equality between a weighted sum over the Gaussian primes and a weighted sum over the zeros. It is an identity, not an inequality or a localization, so it relocates the difficulty (a statement about the zeros becomes an equivalent statement about the primes) without resolving it: GRH, twins, and factoring stay exactly as open on either side.

Glossary

Berry–Keating model. The proposal that a classical Hamiltonian $H = xp$ underlies the Riemann zeros; it reproduces their average counting function but yields no proven operator.

Complex multiplication. The extra symmetry an elliptic curve carries when its endomorphisms exceed the integers — for the lemniscate, multiplication by i from the Gaussian integers; the structure that makes the lemniscate the analogue of the circle.

Conductor. The smallest modulus to which a Dirichlet character is genuinely attached; for the wheel's real characters, a product of a subset of $\{3, 5, 7\}$.

Conductor lattice. The wheel's characters arranged by conductor, forming the Boolean algebra of subsets of $\{3, 5, 7\}$.

Dedekind zeta function. The zeta function of a number field, summed over its ideals; for $\mathbb{Q}(i)$ it factors as $\zeta_{\mathbb{Q}(i)} = \zeta \cdot L(\chi_4)$, encoding how the primes split in $\mathbb{Z}[i]$.

Explicit formula. The identity relating a sum over primes to a sum over the non-trivial zeros of ζ — the bridge between the primes and the zeros.

Frobenius element. The Galois symmetry attached to a prime, recording how it splits in an extension; for $\mathbb{Q}(i)/\mathbb{Q}$ it is the character χ_4 — the identity when p splits, conjugation when p is inert.

Gaussian integers ($\mathbb{Z}[i]$). The numbers $a+bi$ with a, b integers, the integers of $\mathbb{Q}(i)$; a rational prime splits among them when $p \equiv 1 \pmod 4$, stays prime when $p \equiv 3 \pmod 4$, and 2 ramifies.

Generalized Riemann Hypothesis (GRH). The Riemann Hypothesis stated for every Dirichlet L-function at once, not for ζ alone; strictly stronger, and equally open.

GUE. The Gaussian Unitary Ensemble, the random-matrix model whose eigenvalue statistics match those of the Riemann zeros.

Hardy–Littlewood constant. The constant $C_2 = \prod_{p \geq 3}(1 - 1/(p-1)^2) = 0.6601618\ldots$ giving the conjectural asymptotic density of twin primes.

Hecke L-function. The L-function of a Hecke character; in the complex-multiplication setting Hecke proved its analytic continuation and functional equation, yet its Riemann Hypothesis is open exactly as for ζ.

Hilbert–Pólya conjecture. The idea that the nontrivial zeros are the eigenvalues of a self-adjoint operator, which would place them on the critical line; no such operator is known.

Kronecker–Weber theorem. Every abelian extension of $\mathbb{Q}$ lies in a cyclotomic field — the reason the circle's division points capture the simplest arithmetic over $\mathbb{Q}$.

Lemniscate. The figure-eight curve $r^2 = \cos 2\theta$; its division points generate the abelian extensions of $\mathbb{Q}(i)$ by complex multiplication, as roots of unity do for $\mathbb{Q}$.

Lemniscate sine. The function $\mathrm{sl}(u)$ inverting the lemniscatic arc-length integral; doubly periodic on a square lattice and obeying $\mathrm{sl}(iu) = i\,\mathrm{sl}(u)$, the complex multiplication by the Gaussian integers.

Möbius transformation. A conformal motion $z \mapsto (az + b)/(cz + d)$, geometrically a rotation of the sphere under stereographic projection; the integer ones form the modular group. Despite the shared name, unrelated to the Möbius function or the Möbius strip.

One-level density. The averaged behaviour of the lowest zeros across a family of L-functions, used to read off the family's symmetry type.

Parity barrier. The obstruction that sieve methods cannot separate primes from products of two primes, and so cannot, alone, deliver twin primes.

Position and spectral planes. The two planes the explicit formula links: the primes (here, the Gaussian primes) live in position, their L-function zeros live in the spectral s-plane; the formula is the only exact passage between them.

Primon gas. The toy system whose particles are the primes and whose energy levels are $\log n$; it reproduces ζ as a partition function but has the wrong spectrum for the zeros.

Principal character. The trivial Dirichlet character, whose L-function is essentially ζ; the one character whose contribution the explicit formula predicts directly.

Selberg trace formula. An exact identity on a hyperbolic surface equating the Laplacian's spectrum to the lengths of its closed geodesics — the explicit formula made geometric, and the one setting where a zeta's zeros are realized as the eigenvalues of a self-adjoint operator.

Subprime. A composite number coprime to 210 whose smallest prime factor exceeds 7 — the smallest rough numbers the wheel admits.

Symmetry type. The classification of a family of L-functions by which random-matrix ensemble its low zeros mimic: unitary, orthogonal, or symplectic.

Trace formula. An identity equating a sum over a system's spectrum to a sum over its closed orbits; for the zeros, the primes play the role of the orbits.

Twin-admissible corridor. A residue r modulo 210 for which both r and $r+2$ are coprime to 210 — the only classes a twin pair can occupy.

Two-part test. The book's recurring criterion for a candidate Riemann operator: it must have the zeros as its spectrum, and must not be built from them in the first place.

Wheel. The residue classes modulo $210 = 2 \cdot 3 \cdot 5 \cdot 7$ coprime to 210; the organizing structure of both books.

Windowed difference. The contrast between an observable averaged at the zero ordinates and the same observable averaged at the midpoints between them — the book's basic detection statistic.

Bibliography

This is a guide to the sources behind the book, grouped by the part of the story each belongs to. It is a reading list, not a complete record; the aim is to point the reader to the original results named in the text.

The statistics of the zeros

- H. L. Montgomery (1973), the pair-correlation conjecture for the zeros of ζ, and its agreement with the eigenvalue statistics of large random unitary (GUE) matrices.

- Z. Rudnick and P. Sarnak (1996), the n-level correlations of the zeros of principal L-functions, and the statistical independence of distinct L-functions in a family. doi:10.1215/S0012-7094-96-08115-6

- N. M. Katz and P. Sarnak (1999), symmetry types of families of L-functions, and the function-field model in which the zeros are Frobenius eigenvalues and the type follows from monodromy. doi:10.1090/S0273-0979-99-00766-1

- H. Iwaniec, W. Luo and P. Sarnak (2000), one-level densities and the behaviour of low-lying zeros for test functions of restricted support. numdam.org

- A. E. Özlük and C. Snyder (1999), the symplectic one-level density of the quadratic family of Dirichlet L-functions, for test functions of restricted support — the best unconditional evidence over $\mathbb{Q}$ for the symmetry type of Chapter 1. doi:10.4064/aa-91-3-209-228

- M. O. Rubinstein (2001), the n-level density of the same quadratic family, extending Özlük–Snyder. doi:10.1215/S0012-7094-01-10916-2

- Y.-H. He, K.-H. Lee, T. Oliver and A. Pozdnyakov (2022), "murmurations" — the oscillating correlation between Frobenius traces and rank in families of elliptic-curve L-functions, the pattern named in Chapter 6. arXiv:2204.10140

- G. Tang, B. N. Bhaskar, P. Shah and B. Recht (2013), atomic-norm super-resolution of line spectra off the grid, the signal-processing counterpart to the resolution limit of Chapter 5. arXiv:1207.6053

The operator dream (Hilbert–Pólya and after)

- D. Hilbert and G. Pólya (c. 1910s), the conjecture that the zeros are the eigenvalues of a self-adjoint operator (attribution is traditional; no single published source).

- M. V. Berry and J. P. Keating (1999), the xp model and the semiclassical recovery of the smooth counting term. doi:10.1137/S0036144598347497

- M. C. Gutzwiller (1971), the trace formula relating classical periodic orbits to quantum energy levels — the template for reading the primes as orbits. doi:10.1063/1.1665596

- A. Selberg (1956), the trace formula on hyperbolic surfaces — the rigorous analogue behind the orbit picture.

- A. Connes (1999), the trace-formula approach to the Riemann Hypothesis and the positivity criterion equivalent to it. doi:10.1007/s000290050042

- L. de Branges, the approach via Hilbert spaces of entire functions; J. B. Conrey and X.-J. Li (2000), the located obstruction to that approach. doi:10.1155/S1073792800000489

The explicit formula and the five faces

- A. Weil (1952), the explicit formula and the positivity statement equivalent to the Riemann Hypothesis (distinct from the Weil *conjectures* below).

- X.-J. Li (1997), the positivity criterion $\lambda_n \geq 0$ (doi:10.1006/jnth.1997.2137); E. Bombieri and J. Lagarias (1999), the arithmetic form of the λ_n (doi:10.1006/jnth.1999.2392).

One floor up: where the analogue is a theorem

- P. Deligne (1974), the proof of the Weil *conjectures* (and the Ramanujan bound) — the Riemann Hypothesis as a theorem over finite fields. numdam.org

The walls: GRH, the parity barrier, and factoring

- B. Riemann (1859), the founding memoir: $\zeta(s)$, the functional equation, the explicit formula, and the Hypothesis.

- P. G. L. Dirichlet (1837), characters and L-functions, and primes in arithmetic progressions.

- G. H. Hardy and J. E. Littlewood (1923), the circle method and the twin-prime / k-tuple conjectures with the singular series. doi:10.1007/BF02403921

- F. Mertens (1874), the theorems on sums and products over primes (the setting for the twin constant).
- J. R. Chen (1973), Chen's theorem in both forms — every sufficiently large even number is a prime plus a prime or semiprime, and there are infinitely many primes p with $p+2$ a prime or semiprime — the latter, twin form being the one used in Chapter 8, and the state of the art against the parity barrier.
- P. W. Shor (1994), the quantum algorithm for factoring, marking the factoring wall as different in kind from the others. doi:10.1109/SFCS.1994.365700

www.ingramcontent.com/pod-product-compliance
Lightning Source LLC
LaVergne TN
LVHW070856210926
848482LV00016B/948